TOURISM AND REGIONAL DEVELOPMENT

Ashgate Economic Geography Series

Series Editors:
Michael Taylor, Peter Nijkamp and Tom Leinbach

Innovative and stimulating, this quality series enlivens the field of economic geography and regional development, providing key volumes for academic use across a variety of disciplines. Exploring a broad range of interrelated topics, the series enhances our understanding of the dynamics of modern economies in the developed and developing countries, as well as the dynamics of transition economies. It embraces both cutting edge research monographs and strongly themed edited volumes, thus offering significant added value to the field and to the individual topics addressed.

Other titles in the series:

Alternative Currency Movements as a Challenge to Globalisation?
A Case Study of Manchester's Local Currency Networks
Peter North
ISBN 0 7546 4591 6

The New European Rurality
Strategies for Small Firms
Edited by Teresa de Noronha Vaz, Eleanor J. Morgan and Peter Nijkamp
ISBN 0 7546 4536 3

Creativity and Space
Labour and the Restructuring of the German Advertising Industry
Joachim Thiel
ISBN 0 7546 4328 X

Proximity, Distance and Diversity
Issues on Economic Interaction and Local Development
Edited by Arnoud Lagendijk and Päivi Oinas
ISBN 0 7546 4074 4

Foreign Direct Investment and Regional Development
in East Central Europe and the Former Soviet Union
A Collection of Essays in Memory of Professor Francis 'Frank' Carter
Edited by David Turnock
ISBN 0 7546 3248 2

Tourism and Regional Development
New Pathways

Edited by

MARIA GIAOUTZI
National Technical University of Athens, Greece

and

PETER NIJKAMP
Free University, The Netherlands

ASHGATE

Published by
Ashgate Publishing Ltd
Gower House
Croft Road
Aldershot
Hants GU11 3HR
England

Ashgate Publishing Company
Suite 420
101 Cherry Street
Burlington, VT 05401-4405
USA

Ashgate website: http://www.ashgate.com

British Library Cataloguing in Publication Data
Tourism and regional development : new pathways. – (Ashgate
    economic geography series)
    1.Tourism 2.Tourism - Technical innovations 3.Regional
    planning
    I.Giaoutzi, Maria, 1946- II.Nijkamp, Peter
    338.4'791

Library of Congress Cataloging-in-Publication Data
Tourism and regional development : new pathways / edited by Maria Giaoutzi and Peter Nijkamp.
        p. cm. -- (Ashgate economic geography series)
    Includes bibliographical references and index.
    ISBN 0-7546-4746-3
    1. Tourism. 2. Tourism--Economic aspects. 3. Regional planning. I. Giaoutzi,
Maria, 1946- II. Nijkamp, Peter. III. Series.

    G155.A1.T58936 2006
    338.4'79109051--dc22

                                                                    2005031662

ISBN-10: 0 7546 4746 3

Printed and bound in Great Britain by MPG Books Ltd. Bodmin, Cornwall.

Contents

List of Contributors

Coccossis, Harry
Dept. of Planning and Regional Development, University of Thessaly, Greece
E-mail hkok@prd.uth.gr

Colombino , Ugo
Dept. of Economics, University of Turin, Italy
E-mail Ugo.colombino@unito.it

Cracolici, M. Francesca
Facoltà di Economia, University of Palermo, Italy
E-mail francescacracolici@yahoo.it

De Montis, Andrea
Dipartimento di Ingegneria del Territorio, University of Sassari, Italy
E-mail andreadm@uniss.it

Giaoutzi, Maria
Dept. of Geography and Regional Planning, National Technical University of
Athens, Greece
E-mail giaoutsi@central.ntua.gr

Hatzichristos, Thomas
Dept. of Geography and Regional Planning, National Technical University of
Athens, Greece
E-mail thomasx@survey.ntua.gr

Kamann, Dirk-Jan F.
GRIP, University of Groningen, The Netherlands
E-mail kamann@griponpurchasing.com

Leeuwen, Eveline S. van
Dept. of Spatial Economics, Free University of Amsterdam, The Netherlands
E-mail eleeuwen@feweb.vu.nl

Leontidou, Lila
Unit of Geography and European Culture, Hellenic Open University, Greece
E-mail leontidou@aegean.gr

Masurel, Enno
CIMO, Free University of Amsterdam, The Netherlands
E-mail enno.masurel@cimo.vu.nl

Morella, Elvira
Cronkhite Graduate Center, Harvard University, Cambridge, MA, USA
E-mail emorella@hotmail.comworldbank.org

Mourmouris, John C.
International Economic Relations and Development Department, Democritisu
University, Greece
E-mail jomour@eexi.gr

Nese, Annamaria
Dept. of Economics, University of Salermo, Italy
E-mail Annamaria.nese@libero.it

Nijkamp, Peter
Dept. of Spatial Economics, Free University of Amsterdam, The Netherlands
E-mail pnijkamp@feweb.vu.nl

Papakonstantinou, Dimitris
Dept. of Geography and Regional Planning, National Technical University of
Athens, Greece
E-mail dimpap96@central.ntua.gr

Poel, Pauline
ECORYS Nederland BV
Rotterdam, The Netherlands
E-mail paulinepoel@gmail.com

Rietveld, Piet
Dept. of Spatial Economics, Free University of Amsterdam, The Netherlands
E-mail prietveld@feweb.vu.nl

Riganti, Patrizia
School of the Built Environment, University of Nottingham, UK
E-mail Patrizia.Riganti@nottingham.ac.uk

Robinson, Peter
School of Architecture, Planning and Housing, University of KwaZulu-Natal,
Durban, South Africa
E-mail praplan@mweb.co.uk

Stratigea, Anastasia
Dept. of Geography and Regional Planning, National Technical University of
Athens, Greece
E-mail stratige@central.ntua.gr

Strijker, Dirk
Faculty of Spatial Science, University of Groningen, The Netherlands
E-mail d.strijker@eco.rug.nl

Townroe, Peter M.
Sheffield Hallam University, Sheffield, UK
E-mail ptownroe@melton.force9.co.uk

Vliamos, Spyros J.
Department of Economics, University of Thessaly, Greece
E-mail vliamos@uth.gr.

Preface

Tourist visits used to be a rather exceptional activity in the past, which could at best be afforded only once a year during the holiday season. At present, we observe the emergence of the age of mass tourism. Despite a busy lifestyle – or perhaps as a result of a stressful lifestyle – more people than ever before make leisure trips, sometimes for a long time but in many cases just for short periods. Tourist class cabins in planes can accommodate many more passengers than business class cabins. Geographical mobility is to a great extent accounted for by holiday seekers (in many cases even for short weekend trips).

This drastic change in spatial behaviour is caused not only by economic prosperity and our welfare and leisure society but also by the use of the modern information and communication (ICT) sector, which offers: (i) more direct information on interesting places to be visited; (ii) efficient technological tools to organize and book trips; and (iii) proper techniques for communication with friends and relatives through which physical mobility will be enhanced.

Modern mass tourism will inevitably lead to increased competition among tourist destinations, as each tourist region seeks to attract a maximum share from the total stock of tourists. Consequently, tourism has become a key factor in regional development policy. A main challenge of modern regional policy is to market – through the use of ICT – the attractiveness of a certain region in such a way as to generate growth in tourist visits and expenditures. Thus, ICT has become one of the competitive tools in regional tourist policy.

The present book brings together a set of analytical and policy-oriented studies at the interface of tourism, ICT and regional development. These contributions demonstrate convincingly that the modern tourist sector is still an underexplored area which deserves far more profound study. Support for the collection of these studies by the Greek National Tourism Organization is gratefully acknowledged. The editors also wish to thank Patricia Ellman for her careful editing of this manuscript, Vlado Solanic for his skilful formatting of the various chapters, and Ellen Woudstra for her efficient organization of the book project.

Athens, September 2005

Amsterdam, September 2005

Maria Giaoutzi

Peter Nijkamp

Chapter 1

Emerging Trends in Tourism Development in an Open World

Maria Giaoutzi and Peter Nijkamp

1.1 The magic of tourism

In the eyes of many decision makers and politicians tourism has magic potential. It generates income and is based on the indigenous resources of the tourist areas concerned. Tourism has indeed been a rapidly growing sector and a wide-sweeping socio-economic phenomenon with broad economic, social, cultural and environmental consequences. It is likely that tourism will continue to dominate the international scene for many years to come.

Tourism is part of the leisure sector that is rapidly gaining economic importance. The volume of tourist flows at a worldwide level is showing a continued growth path, mainly as a result of increasing incomes and improvement of transport systems (see Pearce, 1981). People are travelling more frequently and over longer distances for leisure purposes. Our world is becoming a global tourist village. Remote destinations are in easy reach and the modern telecommunications sector provides direct information access to such destinations.

Prosser (1994) argues that the dynamics in the choice of tourist destinations is caused by three motives: conspicuous consumption of an elite class; successive class copying, as lower income classes imitate the elite's behaviour, and the expansion of the tourism frontier (or 'pleasure periphery'). Consequently, there are waves of tourism all over the world. This phenomenon can easily be depicted in a diagram of the tourist life cycle that maps out the various stages of tourist demand at a certain holiday destination (see Figure 1.1). This life cycle incorporates clearly distinct classes of tourists who have their own specific behaviour in the various stages of the tourist life cycle, viz. explorers, adventurers off the beaten track, elites, and early mass and mass package tourists.

Over the past 40 years tourism has become a major activity in our society and an increasingly important sector in terms of economic development. It forms an increasing share of discretionary income and often provides new opportunities for upgrading the local environment. Tourism is increasingly regarded as one of the development vehicles of a region, while it is an important growth sector in a country's economy. In both the industrialized and developing world, tourism is often seen as a source of revenues, as a potential for rapid growth, and as an environmentally-benign activity. Whether or not this is true remains to be seen, however.

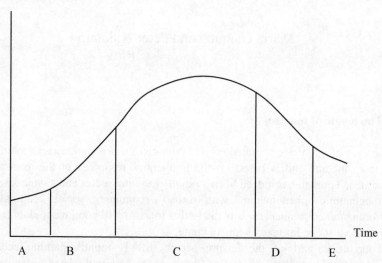

Tourist inflow to
a given destination

A: discovery; B: growing popularity; C: fashion and saturation;
D: fading fashion; E: decline

Figure 1.1 The tourist life cycle

Not only is it evident that tourism is a rapidly growing economic activity, on all continents, in all countries and regions, but it is also now recognized that this new growth sector has many adverse effects on environmental quality. In the context of the worldwide debate on sustainable development, there is also an increasing need for a thorough reflection on sustainable tourism, where the socio-economic interests of the tourist sector are brought into harmony with environmental constraints, now and in the future. Tourism is intricately involved with environmental quality, as it directly affects the natural and human resources, and at the same time is conditioned by the quality of the environment. Much empirical evidence has shown the negative effects of tourism, in particular on the environment. Such a relationship has important implications for policies, management and planning (see Giaoutzi and Nijkamp, 1994).

Tourism is thus a double-edged sword. It may have positive economic impacts on the balance of payments, on employment, on gross income and on production. Moreover, tourism development may be seen as a main instrument for regional development, as it stimulates new economic activities (e.g. construction activities, retail shopping) in a certain area. Nevertheless, because of its complexity, the direct impacts of tourism development on a national or regional economy are hard to assess.

Tourist development poses special problems for environmental resources that are 'exploited' by tourism. The use of such resources has two important consequences. The quantity of available resources diminishes and this, in turn, limits a further increase of tourism. Furthermore, the quality of these resources deteriorates which has a negative influence on the tourist product.

Given the multidimensional dynamics of tourism, it is hard to identify a sustainable development path for the sector, since there are conflicting objectives involved, and the definition of an unambiguous sustainable state for tourism is a thorny question. Müller (1994) has made an attempt to specify sustainable tourism development by using what he called the 'magic pentagon' (see Figure 1.2). This magic pentagon takes for granted that sustainable tourism reflects a state of affairs where economic health, the well-being of the local population, the satisfaction of the visitors/tourists, the protection of the natural resources, and the health of the local culture are all in balance. Any imbalance in this prism means a distortion and will have a negative impact on the benefits of all actors involved.

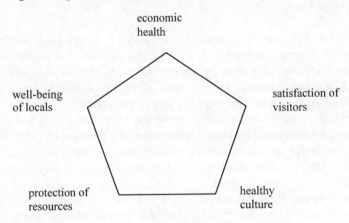

Figure 1.2 The 'magic pentagon' of sustainable tourism

1.2 Impacts of tourism

It goes without saying that tourism induces changes in many areas, not only socio-economic but also environmental. The assessment of such effects is, however, fraught with many difficulties, as the tourist sector comprises a complex set of interlinked activities, such as travel, accommodation, catering, shopping, etc. (see for details Briassoulis, 1995). Consequently, in a strict sense, tourism cannot be considered as a specific sector or industry; it is essentially a complex ramification of economic activities that, in combination, determine the quantity or quality of the tourist product of an area. Hence, given the multi-activity and multi-sectoral nature of tourism, the various sectors that constitute tourism make their own different contribution to the production and consumption of the tourist product.

Firms in the tourist sector are numerous, offering various types of product, each one contributing to the quality of the tourist product. This product is thus a 'packaged' selection of elements that are decisive for the perceived attractiveness of a tourist place. This also means that the determination of a single unambiguous price for the tourist product is almost impossible. Consumers face a multi-product and multi-priced set of amenities, about which they do not have perfect information and free choice, so that they tend to take sub-optimal decisions. And finally, the tourist product is a mix of private and public goods, which complicates the application of purely public or purely private sector policies for the control of its quality and impacts.

An additional complication is that the tourist sector has the typical features of a seasonal activity that may lead to discontinuous economic impacts. In summary, the tourist demand is dynamic, fluctuating over space and time as a result of frequent changes in tourist preferences and marketing policies.

The socio-economic effects of tourism are manifold and can be classified as follows (see Pearce, 1981):

- *Balance of payments*: tourism is essentially an export good which brings in foreign currency, although foreign tourist operators, promotion campaigns abroad, etc. may reduce the net benefits for the balance of payments.
- *Regional development*: tourism also addresses peripheral areas and hence spreads economic activity more evenly over the country.
- *Diversification of the economy*: given the multifaceted nature of the tourist sector, it may help to build up a robust economic development.
- *Income levels*: the income effects of tourism are twofold: building/construction and operations. This may also explain variation in income multipliers (see later).
- *State revenue*: the state earns revenues from tax collection, although it has to be recognized that significant outlays for the infrastructure may also be needed.
- *Employment opportunities*: tourism is rather labour-intensive and also requires much unskilled and semi-skilled labour, which offers great opportunities for less favoured regions.

The extent to which these effects will manifest themselves varies a great deal and is dependent on the stage of the tourist life cycle, local tourist policy strategies and the use of sophisticated communication technology in promotion campaigns. In all cases, the quality of the tourist product offered is decisive for the economic impact on the local and regional economy. In this context, a keen marketing strategy is of the utmost importance, as such a strategy has to ensure the best possible match between the tourist's aspiration level and the opportunity set of the tourist's resources and attributes.

It is a well-known fact that the revenues of the transportation sector – an activity meant to physically transport people and goods – are rather modest. In fact, the transportation sector (interpreted in the above-mentioned limited sense) is, just like the agricultural sector, an infra-marginal economic activity. However, in a modern economy the strategic importance of the transportation sector and the supra-marginal profitability of parts of this sector are dependent on the logistic management of the sector. In other words, it is not the physical movement that is the main source of revenues, but the non-physical organization and coordination of the sector based on modern telecommunication, telematic and logistic services. The latter branch of economic activity does not constitute a low-skilled segment of the labour market, but is determined by highly educated, specially trained and internationally-oriented employees. Thus, logistics and distribution has become more important than material production and transportation.

This observation also applies to the tourist sector. This sector started as a simple, relatively low-skilled segment of the market by offering accommodation and related services to travellers. This traditional picture has drastically changed in recent decades. First, economic prosperity has generated a relatively large share of leisure time and discretionary income, so that more people can now enjoy the benefits and pleasure of international tourism. Second, worldwide mobility has drawn the attention of potential customers to distant and unknown destinations. And, finally, modern telecommunications brings attractive tourist destinations directly into the living rooms of potential travellers. This means that rising welfare and modern information and communication technology (ICT) may be held responsible for the global drive towards mass tourism.

At the same time, a drastic restructuring of the tourist industry itself has taken place. First, a concentration in the sector has occurred: witness the emergence of large-scale international hotel chains. Secondly, as a result of these economies of scale, a further rationalization has taken place, where electronic booking and advanced pay systems have taken over much of the traditional 'handicraft' character of the hotel business. But more importantly, the organization of tourism has come into the hands of a few large-scale tour operators who govern a significant part of the international market. These operators form a critical intermediate segment between demand and supply, in that they do not only organize packages of trips for the traveller, but they increasingly dominate the hotel accommodation market, as well as the tourist transport market. By a keen combination of various opportunities and by using modern ICT as a spearhead, they are in a position to control large parts of the travel agency market and the transport market for tourists. As a consequence, both tourist carriers and hotel owners are becoming increasingly dependent on the vested interests of a highly qualified and technologically well-developed group of tour operators. The question that arises now is: How can regions exploit the opportunities of the modern information and communication sector, without falling into the hands of the monopoly power of international tour operators?

It seems, therefore that a package of measures can be envisaged to nurture indigenous strength and to seek cooperation at the regional level. Elements of such a package are:

- Increase the quality of tourist facilities by addressing in particular the environmental quality of the area.
- Coordinate the information on the supply of tourist facilities (not only hotels, but also culture and nature) at the regional level, so that the region can be perceived as the supplier of a strong package of attractive tourist facilities.
- Invest in sophisticated regional ICTs (for example, electronic booking systems and Internet information on the area via a website page).
- Organize the regional forces so that the international traveller is presented with a uniform tourist image of the area which may favour an inflow of tourism without being dependent on international tourist operators.

1.3 Tourism and ICT

Thus there is a main role for marketing the tourist product by creating new customers through a balanced combination of product, price, distribution and communication services, as such services are often more effective and efficient than other forms of marketing. It is recognized in the tourist sector that markets gain a competitive advantage with improved communication, due to better access to information and distribution, as well as a more proper response to market signals.

Tourism plays a critical role in local economic development in many countries and is an important constituent of the emerging global network society, which is in turn stimulated by the modern ICT sector. The Internet plays an indispensable role in international and national tourism and will most likely become the critical tool for tourism in the future. The introduction of ICT in recent decades has created new opportunities for the tourist attractiveness of remote and peripheral areas, which nowadays also have a virtual access to major centres of tourist origin. This also leads to service competition among tourist facilities in areas of destination, where firms are increasingly involved in global competition (even when they belong to the SME sector).

The past few decades demonstrate a continuously growing trend in modern societies towards long-distance tourism. This trend has been to a large extent the result of:

- a steadily increasing income available for tourist activities;
- an increase of the time available for such purposes;
- greater mobility of people due to the shrinkage of distance resulting from new technological developments in the transportation sector (mainly the airline sector);

- the expansion of the transport system towards new destinations;
- people's changing behavioural patterns and lifestyles connected to travelling, due to increasing internationalization trends in the information era;
- logistic developments of the ICT sector.

The introduction of the various ICT applications related to the tourism sector opens new horizons for the introduction of new tourist services of either existing or new emerging tourist resorts in peripheral areas. As a result, the position of these resorts in the tourist market will be strengthened. The need for an updated network infrastructure, modern telecommunications systems and a skilled labour force is of the utmost importance in the context of these applications.

Such applications are, *inter alia*, focussing on:

- the promotion of tourist destinations through the advertisement of the tourist product in the context of multimedia applications;
- interactive communication between interested parts (tourist destination and the tourist);
- on-line transactions between the tourist destination and the tourist, such as booking, payment, etc.;
- teleworking applications which give the opportunity to combine work with vacations and thus eventually lengthen the duration of leisure time;
- telemedicine applications which encourage aged people to enjoy themselves away from home;
- transport telematics which aim at the more efficient management of the tourist flows, etc.

International tourism follows the patterns of the international division of labour between developed and less developed regions of the world. This division increases the dependence of the less developed countries or regions to a considerable extent, since tourist demand has been largely based upon the cyclic fluctuations of the economies of developed countries. This type of control relationships is becoming even stronger in cases where the tourist market is controlled by foreign travel mechanisms – tourist operators and large hotels.

1.4 Organization of the book

The aim of this book is to explore the present and forthcoming trends in the triangle of tourism, telecommunications, and regional development. More precisely developing patterns of theoretical concerns are presented in combination with methodological and policy perspectives in the thematic area of tourism.

The book consists of three parts. In Part I (Chapters 2–7) 'Tourism, Regional Development and Communications Technology', there are six chapters which make an effort to elaborate on the above-mentioned triangle, which has at its core the ICT developments and their impacts on the tourism sector and regional development.

In Part II (Chapters 8–12) 'Methodological Advances in Tourism Research', a set of five chapters exhibit a broad range of methodological approaches that may provide a sound toolbox for both research and policy analysis in the tourism sector.

Finally, in Part III (Chapters 13–17) 'Policy Strategies on Tourism', a set of five chapters presents a rich package of policies, at the various levels of application, all in the context of the tourism sector.

The first contribution in Part I is Chapter 2, written by Morella, who describes the way information technologies have affected the way travel is marketed and sold, with far-reaching consequences for the tourism sector. A global tourism market has emerged, with new opportunities but also new threats for firms in traditional tourism areas. The design of appropriate polices to support the adoption of modern ICT developments is a great challenge. This chapter reviews the structure of the tourist market and analyzes the possible impacts on the various stakeholders (consumers, suppliers, intermediaries, etc.). Finally, it discusses policies and strategies conducive to a tourism development process relying on modern ICT applications.

In Chapter 3, Kamann and Strijker develop a methodology to create adequate supplier networks named 'reverse network engineering' networks, and combine a top-down with a bottom-up approach. The *top-down* part starts with deriving certain 'concepts' from lifestyles and consumer profiles. It continues with a listing of the products required to produce these concepts and the associated activities and types of actors. Initially, this is performed in *economic space* and subsequently is translated into *geographical space* by taking into account the locational preferences, availability and requirements of activities and actors. The *bottom-up* aspect of the methodology emphasizes the suitability, feasibility and acceptability of developments and/or plans with respect to a particular area and its inhabitants and opinion leaders. The methodology ends with a step where top-down and bottom-up results are matched and screened. The methodology is applied to tourism as an example.

In Chapter 4, Stratigea, Giaoutzi and Nijkamp elaborate on the creation of the virtual organization as a new form of business, defined by its product-market strategy, network structure, information systems and business communication patterns, that is emerging as a response to increasing competition and to the need for the efficient use of resources. They argue that virtual structures and teleworking possibilities imply a greater flexibility for tourism in terms of space, time and structure and enable local communities to compete in the international markets. Thus they focus on the potential role of the virtual organization in local tourism development especially in remote peripheral areas. Policy implications in this respect are explored so that peripheral areas can promote their tourist assets and development.

In Chapter 5, Papakonstantinou examines the conflicts emerging between e-marketplaces and suppliers of the tourism product. His aim is to examine the conflict problem, explore its dimensions and analyze the conditions under which small networks operate. In the first part of this chapter, the problems involved in e-travel business are identified, while in the second, the factors shaping the e-travel business chain are traced. Finally, in the third part, a possible solution is presented in the context of small Business-to-Business closed networks.

In Chapter 6, Stratigea and Giaoutzi focus on the potential role of virtual organization structures in local tourist development, with emphasis on comprehensive tools supporting local tourist development in remote and peripheral areas. First, they present contemporary trends in the tourist sector by putting emphasis on all aspects of the tourist sector affected by ICTs. Secondly, they elaborate on the contribution of virtual organizations in improving the competitive position of tourist SMEs located in peripheral areas. Thirdly, they explore further comprehensive tools based on ICTs, which support local tourist development in peripheral areas. Finally, they draw conclusions and point out further research issues.

In Chapter 7, Leontidou elaborates on the impacts of ICT expansion on movement and hybridity due to tourism and migration, from the vantage point of the cities, islands and coasts of Mediterranean Europe. She notes that the interesting antithesis between Northern and Southern utopias, of urbanism in the Mediterranean and anti-urbanism in Anglo-American cultures, which, in the course of European history, has created different types of tourism and migration, is at present affected profoundly by ICT expansion and postmodern change in consumption patterns, patterns of tourism and migration. The expansion of what is called 'residential tourism', familiar from France, towards the western (Spain) and eastern (Greece) coasts of the Mediterranean, results in new types of migration with an impact on the littoralization observed along the European sunbelt since the 1980s. New opportunities of distance learning, teleworking and communication in general during this period of the digital revolution are affecting population densities in the countryside and the coasts. Trends already evident in case studies from Spain but especially from Greece relate to several reversals observed along the Mediterranean coasts and throughout Europe on most spatial levels: urban/ rural, North/South, East/West, production/consumption-led migration, and others.

Part II of the book provides a set of methodological developments applied to the field of tourism in a modern economy, both locally and nationally.

In Chapter 8, van Leeuwen, Nijkamp and Rietveld seek to compare the economic impact of tourism expenditures in several regions in the world. In many existing studies input-output modelling methods have been used to estimate these economic impacts. Often regional tourist multipliers are derived, which measure the effects of tourist expenditure on the economy of a given area. The tourism expenditure effects can be found, *inter alia*, in regional income, sales and employment figures. On the basis of a broad sample of empirical studies on tourist multipliers, differences among these multipliers are examined *ex post* by means of a meta-analytic approach. The emphasis of the chapter is on both commonalities and contrasts among these study findings. The chapter is both exploratory and explanatory in nature, and aims to map out underlying patterns in the values of the multipliers. While the different regions are characterized by, for example, different population sizes or levels of tourism activities, they also have common characteristics. Using linear regression and rough set analysis as analytical techniques, the relationships between the size of multipliers and background characteristics of the region are revealed.

In Chapter 9, Cracolici and Nijkamp offer an application of Data Envelopment Analysis (DEA) in order to identify and compare the competitive position of Italian provinces in the field of tourism. Since tourism has become a global market, it is of critical importance to identify the relative competitiveness of tourist destination regions. Using several relevant quantitative indicators, this study was able to assess the relative tourist profile of the provinces in Italy.

In Chapter 10, Hatzichristos, Giaoutzi and Mourmouris propose a methodology for the delineation of ecoregions for tourism development in order to enhance the methodological toolbox serving the goal of sustainable tourism. The approach involves integrating fuzzy logic and GIS: on the one hand, GIS provides a powerful set of tools for the input, maintenance and presentation of data, and, on the other hand, fuzzy logic enables the treatment, in the tourism context, of environmental phenomena which are not exact or precise but rather fuzzy. The integration of fuzzy logic into a GIS context increases the analytical capabilities of the proposed methodology for the delineation of ecoregions for tourism development, thus tackling some of the limitations met in the previous delineation approaches.

In Chapter 11, de Montis and Nijkamp provide a novel approach for structuring an evaluation procedure of the territorial tendency towards developing sustainable tourist-based activities. The methodology proposed belongs to the family of multicriteria tools and is developed to integrate the Regime method (Hinloopen and Nijkamp, 1990) with the AHP method (Saaty, 1988). The possible volatility of the results is studied through sensitivity analysis, while the potentials and requirements of an Internet-based evaluation and learning process, grounded in a remote access debate among the stakeholders, are considered as well.

In Chapter 12, Riganti, Nese and Colombino discuss ways of improving the management of cultural heritage sites, by focussing on new forms of involvement and public participation based on the elicitation of public preferences. The problem of an appropriate level of democratic participation needs an *integrated approach*, capable of bridging the practice of urban design, conservation of the built environment, and decision-making support system. This chapter reports results from a survey using *conjoint choice approach* questions to elicit people's preferences for cultural heritage management strategies for an outstanding world heritage site: the Temples of Paestum, in Italy. The potential of the above-mentioned methodologies within the current cultural heritage research scenario is also discussed.

In Part III of the book, the set of chapters presented elaborates on various policy-related aspects of the tourism sector.

In Chapter 13, Poel, Masurel and Nijkamp address the economic importance of a particular class of tourists, viz. those visiting friends and relatives (VFR). After a review of the literature, this chapter offers an empirical case study of VFR tourism with reference to Surinam. The findings show that the economic impacts of tourist visits to friends and relatives are by no means negligible, but represent a heterogeneous spending pattern.

In Chapter 14, Townroe focusses on how SMEs, which are all users of both soft and hard technologies, might more readily benefit from the technologies available to them, but also on how these firms could gain assistance in order to generate their own new technologies. The chapter also elaborates on the channels supporting the SMEs based in their local area (public and semi-public agencies) by providing elements of subsidy and financial support to those small enterprises deemed to be deserving on one criteria or other, normally in terms of an aspect of market failure. In many cases, they also have a training and education role, give support to the acquisition and development of technology, and in this respect, suggest ways of assisting tourist-related businesses, especially those businesses located in lagging regions or sub-regions.

In Chapter 15, Robinson is working on policy issues for the tourist development of the Zululand region. Regional and local planning studies have identified heritage and eco-tourism potential as an opportunity to diversify the local economy. Much of this tourist potential is vested in the town of Ulundi and in the adjacent Makhosini-Ophathe Heritage Park (EOHP). The Park, which is the first to be proclaimed under both nature conservation and heritage legislation, is regarded as the flagship project for tourism in the area. Notwithstanding these opportunities, progress in translating potential into viable tourism products and marketing has been slow. The chapter examines the reasons for this and the obstacles encountered, assesses those aspects of tourism development which are becoming successful, and reflects on whether ICT could be a missing ingredient in the attempt to establish tourism as a pillar of the area's economy.

In Chapter 16, Coccossis aims at presenting the broad range of prospects created by the ICT developments for island regions. Islands often evoke images of pristine environments and cultures untouched by the pressures and frenzy of modern urban living, more like 'small paradises on earth' for living close to nature, places of retreat and leisure, places to dream about and possibly escape to in the future. However, islands may be also seen in negative terms as remote and desolate communities of small, closed and backward societies, places of despair and limited opportunity, lacking resources and modern amenities, places to escape from towards freedom of opportunity and modern easy urban living. In the light of these problems, the chapter presents a set of policies promoting the removal of ambiguities reflected in the conflicts, myths and misconceptions of modern regional development theory and policy.

Finally, in Chapter 17, Vliamos examines the relationship between regional development, environment and the tourist product. The chapter elaborates on this triangular relationship which may lead to the establishment of a system that makes environment an efficient input to the whole production process.

References

Briassoulis, H. (1995), 'The Environmental Internalities of Tourism', in H. Coccossis and P. Nijkamp (eds), *Sustainable Tourism Development*, Avebury, Aldershot, UK, pp. 25–40.

Giaoutzi, M. and P. Nijkamp (1994), *Decision Support Model for Regional Sustainable Development*, Avebury, Aldershot, UK.

Müller, H. (1994), 'The Thorny Path to Sustainable Tourism Development', *Journal of Sustainable Tourism*, 2(3): 106–23.

Pearce, D. (1981), *Tourist Development*, Longman, New York.

Prosser, R. (1994), 'Societal Change and Growth in Alternative Tourism', in E. Carter and G. Lowman (eds), *Ecotourism, a Sustainable Option?*, John Wiley, Chichester, pp. 89–107.

PART I
Tourism, Regional Development and Communications Technology

Chapter 2

Information Technologies and Tourism Development in Developing Markets

Elvira Morella

Information is the life-blood of the travel industry.

Sheldon, 1997

2.1 The structure of the tourism market

The structure of the tourism and travel market is very heterogeneous, with many different players of different size and with various economic performances. These are indicated in Figure 2.1 (Werthner and Klein, 1999), which distinguishes between the supply and the demand side and the respective intermediaries. Only the main players are specified and indicated by nodes, while links mark the interactions and the flow of information that connect them.

Enterprises like hotels, restaurants, etc., but also such actors as cultural and sport events organizers qualify as primary suppliers. On the basis of a functional classification, we find 'big' players such as airlines or railways at the same level as these enterprises.

The supply of basic services is combined in a variety of packages by the Tour Operators (TOs), which thus work as 'product aggregators'. Depending on their size and location, they are distinguished as National Tour Operators (NTOs), Regional Tour Operators (RTOs), and Local Tour Operators (LTOs). Tour operators are often related to primary suppliers through incoming agents, who act on their behalf within a destination.

The final link to consumers is ensured by travel agents, who can be viewed as information brokers, as they provide the final consumer with the relevant information and booking facilities.

Central Reservation Systems/Global Distribution Systems (CRS/GDS), stemming from the airline reservation systems, but now also extended to other means of transportation, provide a link between tour operator systems and travel agents.

The links to governmental bodies are dotted lines in order to indicate that these Destination Marketing and Management Organizations (DMOs) are also often governmental organizations.

The straight line connecting primary suppliers and tourists denotes the possibility for the tourists to choose and book the whole trip through on-line services, typically the Internet.

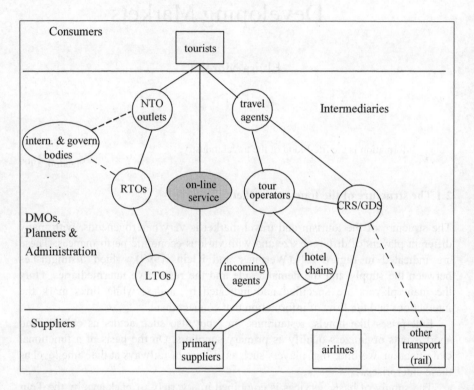

Figure 2.1 The structure of the tourism market

2.2 The tourism market undergoes structural changes

Tourism serves a worldwide and heterogeneous consumer community, with changing needs and buying patterns. Globalization has affected tourism turning it into a real global business, but has also stimulated the development of product variety and of finely targeted product marketing. The demand trend has shifted from mass consumption, which is influenced by product features, especially price, to segmented and individualized consumption, which triggers competition among channels rather than among products and suppliers. In this consumer-driven tourism market, the success of the product depends on the successful extensive targeting of the market and on the development of new distribution channels (Sheldon, 1997).

The main exogenous factors driving the structural changes that the market is experiencing can be summarized as follows:

- *Demographic and social changes*, which affect the countries where tourism flows originate. The progressive ageing of their population increases the portion of tourists belonging to high-income groups and, consequently, the average tourist's willingness-to-pay. This phenomenon has obviously relevant impact on such an income-sensitive activity as tourism. In fact, tourism goods qualify as non-primary within the consumption basket; their consumption is expected more at medium–high levels of wealth, and the supply's variety and quality is strongly determined by the tastes and willingness to pay of consumers.
- *Technology*, which includes information technology, but also any technology that, applied to construction and manufacturing processes, generates improved products/services at decreasing costs, especially in the transport industry.
- *Investments in facilities, infrastructure and equipment*, which determine the possibility to travel and in general the access to tourism products. For instance, where infrastructures are poor and investment capacity remains limited, the practicability of tourism facilities will be seriously constrained.
- *Institutional and legal framework*, which can either impose constraints or provide opportunities: for example, by breaking down entry barriers to travel and tourism, i.e. as in the case of deregulation and privatization policies.
- *Environmental issues*, which are attracting a growing awareness and are putting increasing pressure on suppliers and destinations.
- *Safety*, which constrains the development of specific destinations, especially as a result of the higher sensitiveness to safety requirements that has been developed by many Western countries, from which tourism originates.

2.3 Tourism as an information business

Tourism products cannot be physically displayed or inspected at the point of sale before purchasing. In the tourism market, consumption will necessarily take place in the future and coincide with the production itself. The tourist has, in fact, to travel to the place where the product – the tourism service – is offered, and there the product is consumed at the same time it is produced.

This transforms the relation between demand and supply into a social interaction between the supplier and the consumer, which mainly defines the quality of the product. The tourist takes a leisure decision, which is not only driven by price comparison but is also an emotional choice, often influenced by secondary aspects such as a certain design of the product advertisement. Thus, the appropriate packaging of the tourism product, which is typically a combination of basic products (hotel, travel, etc.), can be crucial. Production, aggregation and distribution are information-intensive and require a major effort in terms of strategic planning.

2.4 Information technologies: The lifeblood of the tourism industry

Information is the raw material for the tourism industry. As few other economic activities require the same degree of generation, gathering, processing, application and dissemination of information, information technologies (ITs) are pivotal for tourism. They incorporate not only software and hardware but also telecommunication systems that enable different players to share information and process it, as well as equipment utilized for the production and provision of services. Besides changing operational practices, ITs also alter competitiveness both at enterprise level and among different regions.

The most significant strategic implication of ITs is the development of comparative advantages that contribute to both reducing costs and enhancing differentiation. For example, the growing use of the Internet for the distribution of tourism products has led to the establishment of an electronic (e-) marketplace that operates constantly, regardless of political boundaries, location and conventional timetable.

ITs help suppliers to identify niche markets; to segment customers on the basis of their different needs; to package and distribute customized products that better match these needs; and finally to sell products at premium prices.

However, sustainable and effective investment in ITs requires monitoring and evaluation at each stage of production and distribution, through, for instance, cost and benefit analysis. Basic prerequisites for initiating investment can be identified, such as:

- re-engineering of the entire business process on the basis of the monitoring and evaluation activity;
- long-term planning strategy;
- top management commitment;
- rational management and marketing strategy;
- use or adoption of appropriate hardware and software;
- skilled human resources, appropriately trained at each level of the hierarchy;
- most important, long-term vision and commitment.

2.5 The challenge for tourism

The typical perishability of tourism products generally induces suppliers to focus on short-term results rather than on longer-term trends and threats, and discourages investment in ITs. Investments are even more heavily constrained in those economies where tourism is dominated by small, uncoordinated and undercapitalized enterprises providing for single service components (accommodation, attractions, transport, lodging, events, tours, etc.).

In addition, the choice to invest in ITs results from a complex decisional process, which has to take into account the interests of all the different actors involved (i.e. consumers, primary suppliers, etc.) and also entails the evaluation of all the possible options in terms of technological solutions (Werthner et al., 1997).

2.5.1 The consumer perspective

From the consumer's point of view, ITs are supposed to facilitate the provision of information, when and where necessary (pre- and/or on-travel information), and the satisfaction of a global variety of needs. Since everybody is a potential traveller and tourist the consumers' community extends worldwide, is immersed in different social and cultural environments, and speaks different languages. Consumers need information before and during their trip and/or vacation: namely, they have a 'dynamic' access to information. And, since they do not necessarily reveal their needs immediately, while making a reservation or at the beginning of the trip, but can change their preferences anytime during the trip, even the 'user identification' can be defined as dynamic. In addition, non-frequent users play a special role within this community, and addressing them poses a major challenge.

Consumers seek information related not only to basic services, such as hotel reservation and accommodation, but also to a wide and customized range of additional services. Cultural events, gastronomy, sport, folk traditions, museums, concerts, etc., can all constitute a significant component of an interesting tour, and the accommodation and means of transport are then defined accordingly. The cultural heritage is probably one of the most relevant 'triggers' for tourism, making the difference between the 'leisure-oriented' travelling, looking for beaches or amusement parks, and the more 'think-oriented' travelling, interested in exploring the richness of different cultures, traditions, and arts and crafts.

In conclusion, the overall quality of services depends on faster and more reliable access to services, wider choice of options, capacity to ensure a tailored package, improved quality of the offered services, and also certified information, which makes users have more confidence in the service they are going to receive.

2.5.2 The primary suppliers' perspective

Suppliers are mainly concerned with the possibility of reaching a wider range of potential users – namely increasing their penetration of the market – and of selecting the right target to address.

Suppliers differ in size, performance, market power, know-how and business interests. In developing and less-developed countries, most enterprises are small and medium-sized enterprises (SMEs), which diverge in quality standards and services but are mainly characterized by limited human resources' skills, especially for what concerns the access and the use of ITs. SMEs working in the tourism sector can be divided into three basic groups on the basis of the activities they perform (Buhalis, 1995), namely:

- SMEs that supply core tourist services, such as accommodation, lodging, site management, as well other enterprises such as retail shops for arts and crafts, information offices and tourist attractions;
- enterprises that participate in the tourism industry, although they are not strictly related to tourism business, such as those providing catering or laundry services to hotels, or local publishers that extend their production to include maps and tourist guides;
- enterprises that would benefit from increased tourism flow to the local area, such as all retail shops in the proximity of a tourist site.

The impact of tourism on a local economy or vice versa the development of the tourism sector in a certain area can be measured by considering the degree to which tourism business expands, until it reaches SMEs in the third group (Buhalis, 1999).

SMEs play a leading role in local economic development because of their capacity to create employment and consequently generate income. SMEs more easily absorb excess labour as their production activities are typically highly labour-intensive and require limited capital inputs. In addition, SMEs are better positioned to generate networks, which are vehicles for know-how dissemination, and therefore facilitate the development of technological bases in poor areas. In both developed and developing economies, there is evidence that clustering and networking of small-scale businesses help raise individual as well as collective competitiveness. Information technologies facilitate this as they help SMEs to establish networks. They provide SMEs with access to a variety of sources of information, including public information, that are of assistance in administrative procedures; they support marketing channels worldwide and enable SMEs to access electronic commerce and benefit from its wider business opportunities.

Information can be grouped into three main categories:

- *external information*: namely, that concerning the environment in which an enterprise operates and on which its commercial opportunities depend. This includes, for example, information about the market, including competitors, customers' willingness to pay, market segmentation; and about public support, including government funding and subsidies;
- *internal information*: generated by the enterprises themselves through their experience, and including know-how, best practice, management procedures, etc.;
- *information sent by the enterprise to the market*: primarily through advertising and marketing, but also including different types of information, such as communications sent to government and public bodies in order to get support.

This classification helps to assess the needs in terms of information in a given situation. External information, especially that concerning the market, such as on prices, price changes, competition, etc., is frequently scarce in less-developed contexts. In this case, entrepreneurs have to appeal to informal networks, as well as having to rely on public support for business development.

The lack of external information is often combined with the total incapacity to develop internal information, and thus generate and disseminate know-how. Enterprises focus on day-to-day management instead of capitalizing on experience to build future capacity. In these contexts they do not have the means to engage in marketing and advertising, and often neither do they receive proper assistance from governments and public institutions.

The proliferation of ITs in developing contexts still suffers from significant constraints. The required investments in IT infrastructure generate debate on their appropriateness as they put a heavy burden on SMEs' resources. Moreover, in order to compete in wider markets, SMEs should be able to market their product together. Consequently, government policies and programmes should help them to pursue agglomeration and develop collective branding.

In conclusion, the use of ITs for tourism development in contexts dominated by SMEs rests on the government's capacity to facilitate investments by the private sector. The efficacy of government initiatives and incentives strongly depends on local communities' involvement and would benefit considerably from NGOs cooperation.

2.5.3 The intermediaries' perspective

Intermediaries act both as product aggregators (tour operators) and information brokers (travel agents). In a phase of advanced application of ITs, the activity of the intermediaries fully depends on computerized workflow procedures. This is understandably not the case in developing contexts, where the tourism sector is driven by SMEs. Moreover, for many destinations, even in a developed context, the most important link is the direct link between consumer and suppliers, traditionally not supported by any electronic means. However, intermediaries, and in particular the tour operators, constitute and will constitute an important channel in product distribution, especially for consumers who do not know the product and/or are not able to use on-line services.

As can be inferred by Figure 2.1, they participate in a complex hierarchical network, dominated by the search for reliable and qualified information, and by the effort to put such information coherently together.

2.5.4 The DMOs' (and Administration and Planning Organizations') perspective

The objective of Destination Management Organizations (DMOs) is to promote a certain area as a tourism destination whilst preserving its social and natural environment, its cultural background and its economy. The tasks of DMOs are manifold, namely:

- strategic planning and destination management, including destination marketing/branding;
- training and education;
- daily operations.

In Figure 2.1 they are associated with Administration and Planning Organizations, which are normally government bodies, since in many countries, and particularly in developing ones, DMOs are genuine governmental institutions. In this case, they earn from taxes applied to tourism activities and do not deal with reservations.

The institutional framework may be completed, and in this case also complicated, by the presence of a third type of organizations, the *local tourist offices*, which may either belong to governmental institutions or be the only organizations covering the functions of DMOs.

Local tourist offices normally use non-computerized information systems, but where they have access to technologically-advanced information they provide local suppliers with most of the support they need, updating them on current trends, the general market situation, and national and international competitors.

Though the specific institutional framework may differ from country to country, nearly all destinations have DMOs. Whether they are government organizations or simple tourism offices, they are in a suitable position to gather information about the local, regional or national tourist product and distribute this information worldwide, as well as to start partnerships with private companies, or set up their own companies.

Public policies can strengthen DMOs' efficiency and effectiveness by supporting them in the use of ITs for improving activities such as:

- monitoring tourist flows, using distributed reservation/accounting systems; assessing customers' satisfaction, designing new tourism packages, planning initiatives and products through information collection, database/recording systems;
- marketing and advertisement;
- provision of on-site information and interpretation services;
- business development.

The implementation of a tourism development programme in a certain country starts from the analysis of the institutional framework regulating the sector, including an assessment of the relations among the organizations mentioned as well as of their respective roles. The governance of the sector is mainly an issue of *negotiation* and *cooperation*, which the introduction of ITs should help to improve in favour of DMOs by helping them to:

- Negotiate and cooperate with the intermediaries, even replacing their role. This, considering the structure depicted in Figure 2.1 (Werther and Klein, 1999), would imply their moving from the left to the right side, and a new market structure as outlined in Figure 2.2. In this structure, the relations with the governmental bodies are traced with a dotted line, since they can change depending on the different institutional framework, while interactions with tour operators and travel agents are ensured by on-line services, as would happen in the best scenario.
- Improve DMOs' control over the variety and quality of tourist services. However, the application of ITs raises two main issues: 1) DMOs are generally not prepared for such an evolution; 2) any sound tourism development programme has to start from the definition of the place occupied by DMOs within the governance framework. This is not always an easy task, as DMOs may play various roles, and often overlap with both government institutions and private enterprises. In case (2), the relations with the private sector pose a sensitive issue to be handled, since the affirmation of DMOs generates competition with private companies that cover the same role, and, consequently, can cause changes in the market structure.

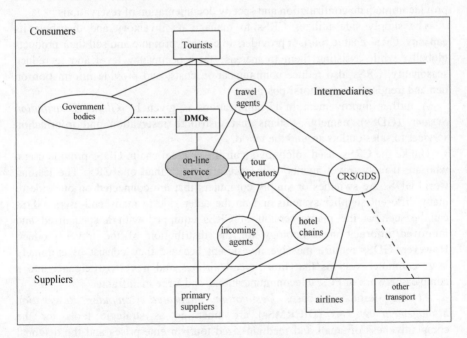

Figure 2.2 Tourism market structure with improved DMOs

2.6 The options in terms of ITs

Computerized networks and electronic distribution are leading to dramatic structural changes within the tourism industry, and are becoming central to the distribution mix and strategy (Buhalis, 2003).

Computer Reservation Systems (CRSs) are seen as the most important catalysts of these changes as they establish a new travel marketing and distribution practice. In its simplest form, a CRS is a database that enables a tourism organization to manage its inventory and makes it accessible to its partners within the distribution channel.

The rapid growth of both tourism demand and supply during recent decades has demonstrated that the tourism industry can only be managed by powerful computerized systems. Airlines were the pioneers of this technology, although international hotel chains and tour operators realized the potential of centralized reservation systems and sought their use.

On the demand side, CRSs ensure consumers can easily access transparent and comparable information on a wide variety of destinations, holiday packages, travel, lodging and leisure services, and on the prices and availability of such services over a certain period. They also enable travellers to book at the 'last minute', and provide immediate confirmation and speedy documentation of reservations.

The supply side utilizes CRSs to manage its inventory and distribute its capacity. CRSs enable tourism providers to control, promote and sell their products globally, while assisting them to increase their occupancy level and to reduce seasonality. CRSs also reduce communication costs, and provide information on demand trends or competitors' position.

A further improvement in IT applications is given by *Global Distribution Systems* (GDSs): namely, systems that distribute reservation and information services to sales outlets around the world.

Unlike the CRSs, used solely by an airline or hotel chain, GDSs provide users, who are usually travel agents, with access to more than one CRS. The leading world GDSs are switches or simply computers that are connected on one side to many different supplier systems and on the other side to many end-users. GDSs can be seen as the macro-version of CRSs equipped with a specialized and improved information technology for the distribution of the travel product. However, GDSs require massive investment because they consist of extremely large computer systems that link several airlines and travel participants into a complex network of PCs, telecommunications, and large mainframes.

At the destination base, *Destination Integrated Computer Reservation Management Systems* (DICRMSs) are emerging as strategic tools for the competitiveness of small and medium-sized tourism enterprises and the diagonal integration of destinations.

These systems aim at improving the information delivery and reservation service for prospective and actual tourists by using widely distributed multimedia presentations.

Through DICRMSs, tourists are informed and enabled to choose and book all services and attractions they want at their destination. While tourists are allowed to tailor the trip to their own needs and preferences, suppliers have at their disposal a strategic tool to plan their offer and integrate with each other. Integration at the destination level leads to branding, marketing and selling the destination as a whole and pushes local economic development.

The systems described so far are based on a type of technology that makes intermediaries, travel agents, tour operators, etc. indispensable in the distribution and marketing of travel and tourism products. New ITs, however, have originated systems in which producers and consumers are directly communicating.

This is what has happened since the integration of a series of high capacity communications channels in the *Internet*.

The Internet and the World Wide Web offer unprecedented opportunities to the tourism industry, as they provide the infrastructure for inexpensive delivery of information on any destination and any type of service. The Internet has solved the problem of marketing remote, peripheral and insular destinations, whose suppliers, namely, SMEs, are able to communicate directly with their prospective customers and differentiate their products according to their needs (Wynne and Berthon, 2001).

The use that suppliers make of the Internet depends on their capacity, their resources, and the technological solutions they can access (van Rekom et al., 1999). The public sector can play a major role to help suppliers, and especially SMEs, to develop capacity in doing business through the Internet. The public sector is asked to provide SMEs with financial support and training; but is also asked to establish a regulative framework that would control the provision and purchase of services by fixing standards, procedures and contract agreements.

2.7 Recommended policy objectives

This analysis has outlined how governments play a leading role in tourism sector development, especially in those economies where supply is dominated by isolated SMEs.

Tourism development strategies are always the yield of cooperation among national and local public administrations, DMOs, and private sector. The government's first task is to organize and foster such collaboration. At national level, this means defining the role of DMOs and providing them with regulations, and appropriate support and funding. DMOs are, in fact, better positioned to: create partnerships with the private sector; transfer government funding to suppliers; and boost their capacity by facilitating the exchange of experience and know-how and by providing for training.

The focus of, and the tasks addressed by DMOs (Werthner et al., 1997) can be summarized as follows:

- *Benchmarking and barriers identification*: The constant benchmarking of the major competitors would help to reveal malpractices and failures in the value chain, as well as to learn about new technological innovations and identify barriers to their adoption.
- *Enabling tourism suppliers, SMEs in particular, to gain capacity to participate in the Digital Economy* (Carter and Bédard, 2001): This task also includes the creation of a training framework to support business innovation in SMEs.
- *Encouraging the vertical cooperation between destinations and intermediaries.*
- *Integrating all the information on a destination*: namely, combining information concerning travel and transport with information on historical and cultural heritage.
- *Pursuing and maintaining research and development activity on a systematic basis* (Frew, 2000): IT tools could also be used for electronic publishing and the dissemination of analysis which is of direct relevance to tourism companies (especially SMEs).

The recommended policies are valid in a broader sense; therefore the need for customized policies is driven by the diffusion of the Internet and the consequent development of e-commerce.

A main task for governments is to design and implement a proper regulatory framework. On one hand, this should liberalize telecommunications and Internet services in order to attract new investments, reduce prices and improve quality of service; on the other hand, it should build consumer confidence, by ensuring privacy, and providing for insurance and protection of payments.

Since not all suppliers may be ready for on-line transactions, destination strategy should include off-line transaction support, local call centers and use of local knowledge.

Again DMOs play a major role in helping SMEs access e-commerce by:

- coordinating the relevant public and private actors;
- integrating new e-tourism channels into traditional and non-traditional distribution channels, and fostering the development of special-interest websites, for example for eco-tourism and cultural tourism;
- promoting destination marketing in key tourism portals, searching gateways, and also listing destinations under as many links as possible;
- conducting research on the application of e-commerce to the tourism sector, in order to provide information on market trends, consumer needs, website use, demographic profiles;
- developing capacity building through training and basic education in relevant fields and conducting awareness campaigns about e-tourism.

In order to do all this, DMOs need to be representative of relevant stakeholders at a local level, institutionally established, and provided with the necessary financial, human and logistical resources.

References

Buhalis, D. (1995), *The Impact of Information Telecommunications Technologies on Tourism Distribution Channels: Implications for the Small and Medium Sized Tourism Enterprises' Strategic Management and Marketing*, University of Surrey, UK.

Buhalis, D. (1999), 'Information Technology for Small and Medium-Sized Tourism Enterprises: Adaptation and Benefits', *Information Technology & Tourism*, 2: 79–95.

Buhalis, D. (2003), *E-Tourism – Information Technology for Strategic Tourism Management*, Prentice Hall, Harlow, UK.

Carter, R. and F. Bédard (2001), *E-Business for Tourism – Practical Guidelines for Tourism Destinations and Businesses*, WTO Business Council, Madrid.

EU DG XXIII (1996), *Applying New Technologies and Information Systems to the Needs of European Tourism*, Working paper, Brussels.

EU DG XXIII (1998), Proceedings of the *Conference on Agenda 2010 for Small Businesses in the 'World's Largest Industry'*, Brussels.

EU DG XXIII (1999), Proceedings of the *Conference on Tourism in the Information Society*, Brussels.

EU DG XXIII (2001), Proceedings of the Conference on *Information Society Technologies (IST) for Tourism*, Brussels.

Eurostat (1999), *Inbound Tourism Flows Rising in Europe*, Theme 4, issue 5, 5 February.

Eurostat (2001), *Domestic Tourism Up in Europe*, Theme 4, issue 16, 20 April.

Eurostat (2002), *Dynamic Regional Tourism*, Theme 4, issue 14, 26 April.

Eurostat (2002), *Stability of Tourism Flows in the European Union*, Theme 4, issue 28, 20 August.

Eurostat (2002), *Tourism and the Environment*, Theme 4, issue 40, 16 November.

Fesenmaier, D.R., A.W. Leppers and J.T. O'Leary (1999), 'Developing a Knowledge-Based Tourism Marketing Information System', *Information Technology & Tourism*, 2: 31–44.

Frew, A.J. (2000), 'Information and Communications Technology Research in the Travel and Tourism Domain: Perspective and Direction', *Journal of Travel Research*, 39(2): 136–45.

Frew, A.J. and P. O'Connor (1999), 'Destination Marketing System Strategies in Scotland and Ireland: An Aproach to Assessment', *Information Technology & Tourism*, 2: 3–13.

Inkpen, G. (1998), *Information Technology for Travel and Tourism*, Addison Wesley Longman, Essex, UK.

OECD (2000), *Measuring the Role of Tourism in OECD Economies: The OECD Manual on Tourism Satellite Accounts and Employment*, Paris.

OECD, co-edited with Commission of the European Communities, United Nations, World Tourism Organization (2001), *Tourism Satellite Account: Recommended Methodological Framework*, Paris.

Proll, B. and W. Retschitzegger (2000), 'Discovering Next Generation Tourism Information Systems: A Tour on TIScover', *Journal of Travel Research*, 39(2): 182–91.

Sheldon, P. (1997), *Tourism Information Technology*, CA International, Wallingford, UK and New York, USA.

Van Rekom, J., W. Teunissen and F. Go (1999), 'Improving the Position of Business Travel Agencies: Coping With the Information Challenge', *Information Technology & Tourism*, 2: 15–29.

Werthner, H. and S. Klein (1999), *Information Technology and Tourism – A Challenging Relationship*, Springer Verlag, Wien and New York.

Werthner, H., F. Nachira, S. Oreste and A. Pollock (1997), 'Information Society Technologies (IST) for Tourism', *Report of the Strategic Advisory Group on the 5th Framework Programme on Information Society Applications for Transport and Associated Services*, Report for the European Union, Brussels.

Wynne, C. and P. Berthon (2001), 'The Impact of the Internet on the Distribution Value Chain: The Case of the South African Tourism Industry', *International Marketing Review*, 18(4): 420–431.

Chapter 3

'Reverse Network Engineering': A Top-down and Bottom-up Approach in the Tourist Market

Dirk-Jan F. Kamann and Dirk Strijker

3.1 Introduction: Networks

3.1.1 The genesis of networks

Companies involved in turnkey projects in less developed areas often face a situation where adequate supplier networks do not exist. Or, that knowledge is lacking about adequate suppliers that potentially may be included into a supply base. Policy makers in the fields of regional industrial development face similar problems. Their colleagues in tourism have to create and/or organise not only suppliers on the input side but also suppliers in most if not *all* parts of the supply chain, including marketing and distribution. Actually, for *policy makers* tourism can be seen as an extreme case of network 'creation', for which reason we will use it as an example.

We may distinguish two different cases: 1) the case where the actor who is the driving force behind setting up such a network is foreign to the area concerned and not too familiar with the area; 2) the case where the driving actor is indigenous to the area where the network has to be created and is familiar with it. Although especially the external actors are advised to start with an analysis derived from Porter's diamond model to check on possible hazards and potentials (for the case of dairy products in Spain see Jurna and Kamann, 2003), local actors too are wise to run a similar pre–check to prevent surprises. Many actors live with a *perception* of what an area can do and the type of actors who are there, which may be either an idealization or a caricature of the area.

The second aspect we mentioned refers to the role and position of *existing* actors. On the one hand, it is possible to complain about 'the absence' of an adequate network in some kind of business. However, there always are actors around who 1) are the initiators who set certain developments into motion; 2) are of potential use in building up the network; or, the opposite, 3) may turn into an important threat for the creation of an effective network. The very use of terms such as 'adequate' or 'effective' here seems to imply that effectiveness is something of an objective measurement. It suggests there is something like a 'Pareto optimal' network. In real life, different actors will of course have different perceptions of 'the effective' network, something we will have to take into account. Apart from this aspect of the role and position of existing actors, we have to ask with many activities in the network *in statu nasciendi* whether an existing group of initiating actors – usually the 'we' in the process – have to perform certain or all activities or that an existing actor that 'we' fear should actually perform those activities in order to be more 'effective' in, for instance, reaching consumers. Given the power of tour operators, many other – less powerful – actors do not want them to get involved out of fear that 'they' may take over. However, by excluding the feared and powerful actors, the result may be that certain activities are either impossible to perform or are highly ineffective. Therefore, a simple rule seems to apply here: *if* the market you are going to operate in is characterized by a certain distribution of power, it is better to play the game *recognizing* that power and therefore following the rules of that market. Then, if that rule is considered to be desirable, actors can build up countervailing power through collective acting, monopolising supply or other forms of collusion. They can change their position vis-à-vis those powerful players and deal with them in such a way that, in the end, a win–win situation occurs. Ignoring them may give the satisfaction that 'they' were unable to dominate, but on the other hand, may put the whole enterprise into jeopardy because of lack of, for instance, customers.

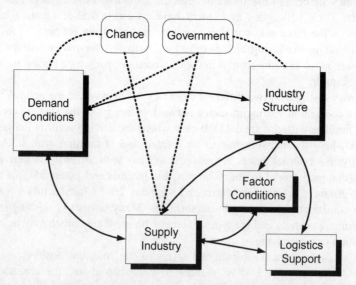

Figure 3.1 The ex-ante analysis derived from Porter

3.1.2 The network approach in general

After this brief discussion about the creation of networks, we will briefly discuss the network approach used in this contribution (see Wasserman and Faust, 1994). This is a synthesis of ideas and concepts of various scientific communities. It combines the actors/activities/resources concept of the IMP-group (Axelsson and Easton, 1992), with other paradigms such as the Transaction Cost Economics – especially the notions of uncertainty, frequency and asset specificity – (Williamson, 1975), the notion of 'Milieu' of the GREMI-group (Camagni, 1991), the 'missing links' concepts of the physical networks of the NECTAR-group (Button et al., 1998), the social interaction and Resource Dependence orientation of the social scientists (Pfeffer & Salancik, 1978) and, finally, the Resource Based View of the marketers (Wernerfelt, 1984). This synthesized approach therewith combines various dimensions: the economic, social-cognitive, spatial and time dimensions of networks. Up to now, the methodology to visualize these dimensions was used to study *existing* networks where the actors already involved are visualized and analyzed together with their activities, their relations and the distribution of resources. Existing flows of goods, services, information, and capital between the actors are taken into account and the nature and extent of dependencies, lobbies and strategies are derived from the analysis. From the analysis, perceived deficiencies in the functioning of the network are derived and these form the basis for policies. The methodology developed and used – the *Triple Plus methodology* – is based on a layered model, where in each layer different techniques are used to show certain aspects of network behaviour. The combined outputs of the various layers enable triangulation of the network analyzed to increase the sharpness of the picture we want to have of the network, the strategic behaviour of its actors and the nature of their relationships.

3.1.3 Reverse network engineering

In contrast to studies of existing networks (see Kamann and Strijker, 1991), the prime goal of this contribution is to give a methodology to *develop* networks that do not yet exist or are only present to a limited extent. Therefore, we have to reverse the order in the manner of *reverse engineering*. We start at the demand side by analyzing the different types of lifestyles of potential target groups and the type of *concepts* in which they are interested. The next questions are in turn: 1) Which products make up the concepts? 2) Which activities produce those products? And only *then*, 3) Which (type of) actors perform those activities? We will spend some more time later on discussing the question: Which products? Having solved that question, we will discuss the technique which enables us to draw up the network of all actors, required – with their activities and required resources – to produce those products *in economic space*. After this, we deal with possible opportunities for economies of scope and discuss the need for a check on possible conflicts between activities and concepts. *Territorial* space will be added in the following section, with the locational requirements and preferences of activities. The actual network for a particular area resulting from this exercise is discussed, together with the way the 'top-down' method described is matched with a 'bottom-up' approach taking the feasibility, suitability and acceptability of policies and plans for an area into account. We continue with a section describing the process of final 'matching and screening', including the scope and conflict issue. After that, the role of a 'visible hand' co-ordinating the process is described. Finally, we will describe a case from the tourist industry where a network was created as a deliberate policy to generate work and incomes in a peripheral area, after which we will summarize the whole process.

3.2 Concepts and products

3.2.1 From market side to creating sourcing potential

The usual textbooks on marketing (Kotler, 1991) teach that any successful organization should be market- and consumer-oriented. Recent research findings show that, as well as the significant relationship between customer orientation and company performance, the relationship between a well-engineered and -maintained supplier network also shows a significant relationship with company performance (Tan et al., 1998). Hence, companies not only should worry about either their customer orientation *or* their supplier orientation but also should be aware that it is the right way of linking up these two that makes the difference. In line with this philosophy, we *start* at the market side and ask which target groups we should focus on. What kind of demand can be expected? And how do we actually *reach* those people – marketing channels, distribution? Only when we know all this can we start building the network that enables the class of actors participating in it to cater for the customer needs (see Borgstein et al., 1997).

The implication for our case of tourism is that when designing strategies for regional development with tourism as a vehicle, any agency involved in this process should wonder what buyers – consumers and industrial buyers – actually want instead of producing something that from the producer's point of view may seem to be marvellous. We could even go one step further by stating that, rather than the physical properties of the product *per se,* it actually is the *concept* people buy, linked to a certain functional need. For instance, some people do not drink 'just' whisky, they also want to consume the exclusive world of caviar, diamonds and sparkling crystal glasses, and imagine themselves part of that world when they drink Glenfiddich. Others associate themselves with the world depicted by Jameson whiskey, while again others live in the Martini world. People tend to select certain products not because they are so much better, but because of the 'image', the 'concept' they represent.[1] We can translate this into tourism and rural areas. 'Living in the countryside' as a total concept means quietness, space, physical activities, the feeling of belonging to a community, small and human scale, a sparsely populated area with low-density building, fresh air, and a healthy and peaceful environment.

Concepts to which people attach a value are dependent on – or are typical for – certain *lifestyles* and *consumer patterns*. Concepts offer people what they ask for in an *im*material sense, apart from the physical product. Therefore, when we want to sell tourist products it is not so much the physical product *per se* that counts among customers in deciding to buy it, but whether it fits the concept the customer wants, that is, whether it appeals to the consumer's lifestyle. Proper marketing may increase the awareness among members of the potential target group that the concept offered is actually part of a lifestyle that appeals to them. Because of this we conclude that it is the concept that has to be sold, including the immaterial features.

Let us take as an example a fishing village with a fish restaurant serving relatively straightforward but high quality food. The concept sold is 'relaxed eating real fresh and honest stuff' and this contains aspects such as good, fresh food, nice view of the sea and/or harbour, plenty of parking space, honest and friendly staff with a personal touch and a relaxed unhurried dinner. The physical products required to sell this concept are the building, the ingredients of the meals, properly trained staff and a parking lot.

Selling a concept can be done in two different ways: 1) as package deal – often purchased in advance; and 2) as separate items, purchased immediately before the moment when consumption is considered to be desirable. The first way provides full arrangements. The customers buy an all-in arrangement, including transport, excursions, hotel and dinners. The selling agency in this case could be seen as a *main supplier*, with the individual producers of the products making up the arrangement – transport, dinner, etc. – the second-tier suppliers. The second way, where the customer will be just attracted to the area based on the knowledge that the concept – and its ingredients – is available puts the producers in different positions. On arrival in the area, the customer may want to make his own choice of, for example, restaurants, doing his own booking, providing his own transport, and

---

[1]   At the same time, we find that people with certain lifestyles express their lifestyle in certain purchasing behaviour which expresses their self-image – clothing during office hours (Boss, Armani), clothing for going to the theatre (Versace), and clothing during collective outings (Calvin Klein) – while opting for different purchasing behaviour for primary goods (Aldi, K–Mart) and clothing to be used at home. In other words, given normal budget constraints, people do not conform in *all* their shopping behaviour and procurement to their lifestyles. Therefore, when an area decides to produce a particular concept associated with a particular lifestyle, this concept has to be marketed with the usual marketing techniques to make the potential buyers aware of their cravings and the solution to fulfil them on the one hand, and of the unique selling points of the areas concerned to fulfil those cravings, on the other.

by doing so opting for more freedom to choose his arrival and departure time. Of course, he may also use the receptionist of the hotel where he is residing to make certain bookings and reservations and will rely on him or her to recommend suitable places. This freedom to determine the customer's time management is likely to be an important part of a particular lifestyle.[2]

Therefore: different concepts – package trips versus individual choice – may require the same products or, put the other way around, the same products may be used for different concepts. This will open the potential for economies of scope for the producers of these products, as we will see later on.

3.2.2 First-, second- and third-tier activities

For each concept, certain physical products have to be produced that are the ingredients for those concepts. In our example, the fish, vegetables, energy, labour and capital to start the business can be described as the *first-tier products*. In order to produce these first-tier products, certain *second-tier* activities are required, either as backward linkages or as forward linkages. Examples are catching the fish, transportation of the fish, bringing in fresh vegetables, transporting the customers who may come from their hotels or may want to return there, cleaning the restaurant, laundering bed and table linen, maintaining the building. In their turn, these activities may well be supported by yet another round of activities, a *third tier*. Examples are the wharf and supply shop for ship maintenance, companies that repair transport vehicles, laundry machines and telecommunication systems, and the petrol pump. Therefore, we can use input/output type multiplier linkages to trace all activities related to the restaurant in the example.

When we deal with concepts that are broader – and vaguer – like the concept 'relaxed quietness' it may be that this concept not only asks for hotels but also for tennis courts, boat rental, some yachting, horse riding, some facilities in case of bad weather, and so on. Again, these 'products' in turn need suppliers, maintenance, repair, cleaning, transport, and so forth, that provide the necessary inputs. Hence, each concept implies a series of first-tier, second-tier and third-tier products.

---

[2]   In fact, the very existence of consumers of 'package deals' in a locality may deter other types of consumers who detest such things as being non-compatible with their lifestyle. They may detest organized tourism and demand immediate availability of tourist items without waiting in queues or other types of congestion. The ultimate implication of this conflict may be to split up an area into a sub-area catering for mass consumers and a sub-area catering for individualists, or make a choice for either of the concepts when splitting up is impossible.

3.2.3 Summarizing step 1

To summarise, the first step in the process contains the following four ingredients: 1) derive the concepts from the lifestyles and consumer patterns of the target group; 2) give a first estimate of the size of the target group and the volume of demand in terms of estimated buyers and demand; 3) derive the products and components that make up the concepts; and finally 4) make an inventory of the activities required to produce those products.

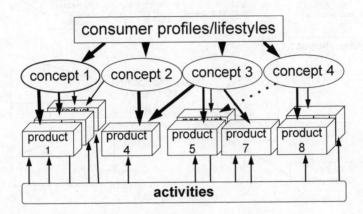

Figure 3.2   Lifestyle, concepts and products

3.3 The filière

In order to produce all *products* – of whatever tier – required for certain concepts, certain *activities* are required, which are performed by certain types of *actors*, who control certain *resources*. This brings us back to our networks. Starting with each individual concept – to be merged later on into a picture of all concepts together – the *second step* is to draw up the network of all actors required with their activities and required resources to produce certain types of products. In the first stage of the process, we only refer to *types of products* at a general level: currency exchange facilities, dinners, bus transportation, overnight stays, laundry services, etc. This type of network with general products is named the *initial filière* (Figure 3.3).

Each box in the filière of Figure 3.3 with the type of *products* shown in fact has a shadow box, which indicates the types of *activities* producing the type of product mentioned in the box. Again, these activities are in general terms: banking, providing dinners, transporting vegetables, fish, people, etc. Finally, the shadow boxes of general types of activities also have their own shadow box: the actors. Actors are taken as a class of certain types of actors, performing the associated activities which produce the types of products in the original box: banks, restaurants, bus companies, hotels, etc. Figure 3.4 gives a detail of Figure 3.3 with the shadow boxes.

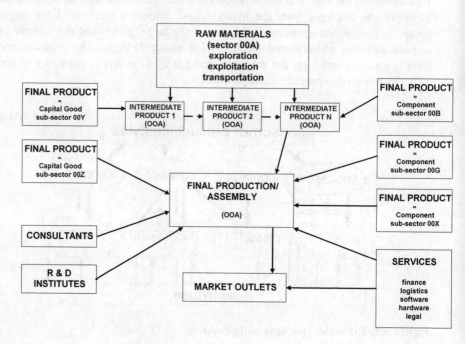

Figure 3.3   The general principle of the filière
Source: Kamann, 1988, p. 60.

The filière as shown in Figure 3.4 is more than the conventional 'production column' or 'supply chain' in the strict sense: it transcends the industry borders. The filière was originally developed as a tool to make an inventory of all existing activities and actors involved in order to *improve* the entire network in respect of quality and completeness. The philosophy behind the filière is that the strength and competitiveness of the actors represented as a group – the network – depends on the performance of the weakest participant. By drawing up and analyzing the filière, the various types of actors could be traced and analyzed with regard to their value and contribution to the network. In addition to this, products and activities perceived to be *missing* from the filière and the actors producing them could be indicated.

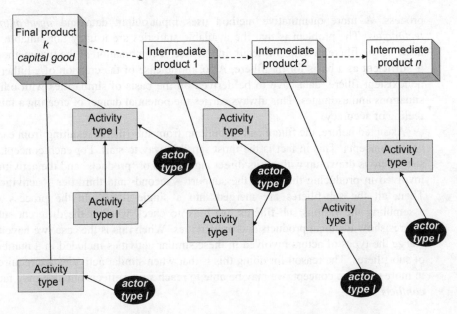

Figure 3.4   The initial filière with types of products, activities and actors

3.3.1 Separation of production, assembly, sales and final use

Traditionally, a single economic and physical unit – one company at one site – performed the production of intermediate products and final assembly. Today, we find an economic and geographical disintegration in this respect. At geographically different locations, legally independent actors produce intermediate products and components that are ultimately assembled in a particular point in economic and geographical space. In our case of tourism, we also may find that the marketing, ticketing and a large proportion of the sales of products – services – consumed in the tourist areas is performed in the areas of origin of the consuming tourists. Hence, purchasing a tourist good may well occur in an area thousands of miles away from the tourist area where final production and consumption take place.

The filière also includes relations with suppliers of capital goods, required to transport, produce, assemble or process inputs and suppliers of specialized services. Because of this, we find that, where the supply chain and production column tend to be taken as activities *within* a single industry sector, the filière transcends the sector borders and includes activities from various industries.

Various techniques can be used to draw up the filière. First of all a *flow diagram* enables the researcher to describe all activities and actors in a qualitative sense. The information required to draw this up can be obtained from interviews and literature. Drawing up the filière this way is usually an iterative process, where new activities are added and existing ones more precisely described during the process. A more quantitative method uses input/output data and *input/output techniques*. The problem is that the available statistics are usually not suitable to draw up the filière in terms of detail. In practice, a combination of the two methods is sufficient as a basis for the filière. Also, in the case of the creation of a hitherto nonexistent filière, data have to be derived on the basis of similarities with other situations and estimates. This always carries the potential danger of creating a false feeling of accuracy.

As stated before, the filière can be built up from sub-filières resulting from each of the concepts. This in fact is the most sensible way to start. For each concept, a sub-filière is drawn up with all its direct ingredients or 'products' and the activities involved in producing these and the support – second- and third-tier – activities. Then, all the sub-filières are merged into a single filière. In the process of assembling the various sub-filières, we have to check whether the different sub-filières show identical products and/or activities. When this is the case, we have to *merge* the types of actors involved in these similar activities included in a number of sub-filières. The reason for doing this is that when similar activities are required in more than one concept, we may be able to reach *scope* effects but also may face *conflicts*.

3.4 Economies of scope versus conflicts

3.4.1 Potential for scope effects

In order to be economically feasible, activities have to reach a certain minimum threshold in turnover. It may happen that the volume of a certain product required by one concept is insufficient. However, when activities provide products that fit in the supply chains of more than one concept, economies of scope can be created. For instance, a booking/ticketing office that arranges farmhouse dinners can also arrange tickets for boat trips or the tennis courts. Similarly, the maintenance and repair shop that repairs the taxis, can also repair rental cars, outboard engines and mountain bikes. It does mean, however, that actors have to be open for scope synergies instead of pure scale effects.

It should be taken into account that producing products or services for different concepts implies that the way of production is not always identical for each concept. Renting boats for touring and sightseeing requires that the customers receive maps with suggestions for sightseeing, and means that the renting-out should be done with a 'personal touch'. Renting boats to people who want to do some active fishing requires the availability of angling equipment for rent for those who go fishing for just one day. This type of customer also requires efficient renting and early opening hours. Eventually, the production for different concepts can create certain conflicts: 'personal touch' and 'efficient renting' can be conflicting. Real fishermen know how to deal with a small boat and do not require time-consuming explanations. They just want to go fishing.

Not only the *production* for two or more concepts can cause conflicts, also the *consumption* can be conflicting. The sightseeing motoring boats are usually negatively viewed by fishermen who take any motorboat or other moving object as a threat to their fisherman's luck. This problem may have to be 'regulated' either by designating certain areas for fishing and 'touring' or by using more hidden persuasive techniques which involve creating and indicating certain areas as attractive for the touring boat renters, which leaves the fishermen a peaceful place in certain areas known and communicated to be favourite among fishermen.

In general, *'scale'* obtained and required is a dependent variable determined by the technology applied and organization. It is influenced by the nature of the products required by concepts and the production processes involved in producing those products. Therefore, when designing the filière that produces concepts, scope effects have to be included to obtain sufficient volume for the actors and their activities involved. This means that more than one concept may have to be combined, 'sourced' by shared actors.

A practical way to visualize and work out the scope effect between actors is derived from Porter's (1985) *value system* with the *value chains* for each actor. By using this instrument, it can be illustrated that actors have the following options: 1) actors can *pool* operations in order to obtain minimum scale effects and/or increase efficiency and effectiveness; the pooled operations can be performed by any of the actors or by a specialized new jointly owned entity; 2) actors can mutually out-source products; two actors may decide that instead of each producing two products with insufficient scale effects, they each may focus on one of the two products and exchange them; this way both products are produced with scale effects; and 3) actors can choose to perform *part* of the operations themselves and *pool* other parts: for instance, incoming logistics, distribution, sales, and so on.

3.4.2 Possible conflicts

In order to create scope effects actors may produce ingredients for two – or more – concepts. However, *conflicts* may occur when two concepts are combined that in the perception of the customer cannot coexist in the same area. 'Mass recreation' and 'peace and quietness' for instance are bound to raise conflicts. This point has to be taken into account when the actual translation into geographical space and actual areas takes place. The existence of potential conflicts does not imply that the combination of concepts is impossible in *economic space*. Therefore, at this stage we can only *mark* the actors/activities involved as potentially causing a conflict. It becomes relevant at the final stage, where we look at an actual area with an actual selection of specific concepts in a territorial setting. It may imply that some actors may still be able to reach scope effects in supplying certain ingredients for certain concepts *but should not be located in the area of consumption*. Transportation is a likely solution to overcome this geographical separation between the location of production and the location of consumption or *further processing*.

Problems also can be overcome by the use of a 'two-doors system', for example, the exploitation of two physically separated restaurants – each focussing on a different concept or segment – in one building, using one and the same kitchen.

3.4.3 Step 2 summarized

The second step, therefore, can be summarized in the following way: 1) draw up a filière which is the total amalgamated version of all sub-filières derived from each concept; the filière is drawn in economic space and describes types of products, types of activities and types of actors; for the time being, geographical references are not included; 2) trace possible opportunities for economies of scope where actors are involved in the supply to more than one concept; use Porter's value chain and value system where appropriate; 3) check on possible conflicts between concepts that are linked through scope-obtaining actors, and mark the activities involved as 'potentially conflicting when within one territory'.

3.5 Space: Locational requirements

3.5.1 From economic space towards geo-space

The reverse engineering that results in a filière producing one or more concepts starts off in *economic space*. Hence, both Step 1 (formulating the ingredients that are required with the activities involved) and Step 2 (tracing and mapping the corresponding types of actors) do not yet have a geographical dimension. The first two steps are performed *irrespective* of the question concerning *where* activities have to take place.

In Step 3, the geographical dimension is added. That implies that activities and actors are analyzed for their locational preferences and restrictions. For instance, some activities have to take place in the market area where the target group lives: the potential customers/tourists. Other activities have to take place in the area where the consumers/tourists go to consume their 'concept': the tourist area. Another type of activities may have to be performed in the metropolitan area (centre location) of the country in which the tourist area is located. In Greece for instance, Athens would be the obvious metropolitan example. Marketing and sales usually are best performed in the area where the demand is: the origin countries and areas of the potential tourists. There, the best way to appeal to certain lifestyles can be tested and an all-inclusive leisure week, activity week, tracking tour, a boat trip, etc., can be promoted, sold and arranged, tickets can be sold and reservations can be made.[3] The preparations and home-base marketing – the analysis of the 'producing' areas and so on – are usually performed by companies in the national metropolitan area mentioned since, in most countries the marketing agencies are located there together with all the headoffices of manufacturing and service companies. The actual boat trip has to be provided in the destination area of the tourists. Of course, some activities (especially support activities like, for instance, administrative back office work) are relatively 'footloose' and can be provided from anywhere: market area, tourist destination area, or elsewhere. These footloose activities are usually best performed at a location, which is the location optimum, balancing minimum distance to markets with minimum distance to resources and technical scale effects. In all cases, proper telecommunications is a *vital prerequisite* for a properly operating filière. When this precondition is met, these activities can be located in the more peripheral areas as part of regional and/or rural development programmes.

Because of this, it has to be decided for each product (and the activity involved to produce it) whether there are any *locational restrictions* or *preferences*. This element is added to each type of activity and, therefore, to each type of actor when the filière is 'translated' from economic space into geographic space. Figure 3.5 gives an example of this *third step* in the process.

3.5.2 Scope effects or conflict: Maintain the markers

In Step 3, which deals with the process of adding locational requirements to the filière, we have to remember the markers on activities with potential scope effects and/or possible conflicts that were installed in Step 2. These markers are necessary in obtaining economies of scale and scope by combining filières that produce different concepts but share the same types of activities, resources and actors. In the same context, we have to incorporate the *conflict* element mentioned before. It may happen, of course, that the locational preference or restriction of an activity that was marked as potentially causing a conflict with a certain concept, geographically separates this activity already from that concept, which solves the problem. Activities marked as potentially causing a conflict that in Step 3 are situated in the destination area where the concepts are consumed keep their marker.

3.5.3 Step 3 summarized

Step 3, therefore, can be summarized as follows: 1) add the locational preferences and restrictions to the activities and types of actors in the filière; 2) maintain the potential scope effect of combining concepts in producing activities; if these concepts involved are separated in geographical space indicate the possibilities to

---

[3]   Most of the internationally operating tour operators have segmented the market into 'concepts'. Proper contact with the major – and powerful – international tour operators is a prerequisite here.

overcome distance through transport and telecommunications and indicate which concept is more likely to suffer from the presence of activities; 3) maintain the markers for potential conflict activities that do not harmonize with certain concepts when they end up in the same area.

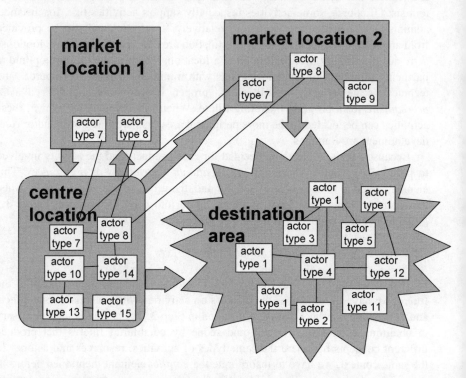

Figure 3.5   Adding locational preferences in geo-space (Step 3)

3.6 The actual network

In Step 4, the result of Step 3 is translated into a *network* for a particular area with named actors. This step has two aspects: a *top-down* aspect and a *bottom-up* aspect.

3.6.1 The top-down aspect of Step 4

The top-down aspect of Step 4 starts with the selection of an area and then goes into the selection of one or more concepts that seem to be suitable, feasible and acceptable for the selected area to produce. The answer to what is suitable, feasible and acceptable will actually be found as part of the bottom-up aspect. Here the process becomes iterative, in the sense that *bottom-up* information about the actual actors present in the area, their activities, and the characteristics of the area is used to select the concepts to be produced by the area. In the resulting network, actual company names of actors replace the general category names in the filière. When actual companies who fit in certain categories of the filière are missing, the category name remains. At the end of the top-down aspect of Step 4 we have an indication of which activities can be produced by actors present in the area, and which activities are still missing because the type of actors needed are not there.

3.6.2 The bottom-up aspect of Step 4

The bottom-up aspect of this step, which feeds into the top-down aspect starts with information about the actual actors and their activities in the area. The Triple Plus methodology had proved to be able to visualize and describe these actors (Kamann, 1988, 2003). The results of this part of the analysis are used to select concepts to be produced by the area. Three aspects play an important role here: feasibility; suitability; and acceptability (Johnson and Scholes, 1989, p. 170).

Feasibility relates to the question of whether any plan proposed can be implemented. Are sufficient resources available to finance the plans? Are the existing actors willing to make and capable of implementing the changes required? And are they qualified to provide the activities specified? Can the market position actually be reached and the concept be sold to attract sufficient tourists to the area? Are there any competitive reactions expected that might upset the fulfilment of the plan? Can the activities and their actors in the area efficiently and effectively be integrated in the proposed network including the 'off-shore' actors in the centre area and the market areas? Can transportation and telecommunication with proper materials management for the entire network be obtained?

The *suitability* of an area should focus on the question of whether the area's strengths are capitalized sufficiently, and whether weaknesses found in the area are being dealt with. To analyze the suitability of an area, one has to make a listing of its strengths, in particular those that give it a distinct competitive advantage compared with other areas. In addition, the weaker points of the area also have to be listed and so too the major opportunities and threats which the area faces. An important element is formed by the shared belief sets of the inhabitants and their leaders; and their objectives – implicitly or explicitly formulated – which influence policies.

Suitability is important when evaluating a proposed development plan to promote certain activities in the area or 'discourage' other activities to produce certain concepts. The strategy proposed should deal sufficiently with weaknesses and exploits the strengths of the area in dealing with the opportunities of the market. It should be in accordance with the objectives of the inhabitants and their formal and informal leaders and representatives and fulfil their expectations.

Acceptability deals with the question of whether the final result that is envisaged by the plan is acceptable to major stakeholders. Questions raised here deal with the perceived financial risk of actors, possible changes in the function and status of actors in the area, the relationship of local actors with outside stakeholders and environmental changes, such as more noise, more people, more activities, and so on.

Eventually this can imply that sometimes cross-subsidies are required in order to convince all participants. This can especially be the case when certain public goods are essential elements of the concepts. In certain cases, this will imply that owners of restaurants, bars or boat rentals should subsidize the owners of public goods.

An important element of the bottom-up part is to find out what all the local actors actually *want*. As trivial as this may sound, in practice we find many cases where learned researchers claim they 'know' what people want, while communications with these local actors reveal quite some differences. We suggest using soft systems methodology (Checkland and Scholes, 1990) and related methods in order to: 1) settle what people want; 2) coach them in finding a common, acceptable strategy. If necessary, this can be combined with change in management tools (Flood and Jackson, 1991).

3.7 Matching and screening

Since we are dealing with an actual area and concepts that actually will be provided in that area, the potential for scope effects has to be explored in this stage, and the possibility of conflicts has to be checked. If conflicts arise, solutions including spatial separation have to be sought while still obtaining scope effects where possible. Proper logistics will most likely enable this.

We also have to include an element, which can best be described as the *negative synergy* effect. This *restrictive* element occurs when activities in an area do not fit the perception of the customers that a particular concept belongs there. Industrial types of agriculture usually are not perceived as 'rural', and customers visiting an area in order to enjoy the rural elements will be disturbed by such unwanted activities. Heavy industry looming over a picturesque village is of course damaging almost by definition. Also, while some activities in small-scale enterprises may be acceptable, the same activity in large-scale enterprises or in many enterprises is not acceptable. One piggery or garage may be acceptable while a whole range of piggeries with their smell or garages with their noise are not acceptable. It is hard to decide on objective grounds what is acceptable or not. Therefore, acceptability should be seen from the viewpoint of the local inhabitants and policy makers, but even when these consider an activity acceptable it still may not convince the *customers* – the tourists – that the area is meeting their standards. Therefore, what is acceptable or not depends ultimately on the perception of the consumers.

The result of the matching and screening process in Step 4 for a particular area therefore has to answer the following questions: 1) Does the resulting planned network create sufficient positive synergy through economies of scope? 2) Are there are any activities that may conflict with the image associated with any concept produced in the area?

The result of Step 4 will be a network that is *required* and *allowed* in an actual area to produce certain concepts with a vector of activities of actors who produce certain products and who are linked up to certain described outside actors with specified activities and products.

3.8 The organizer, facilitator or director

The screening and matching described above may suggest an 'invisible hand' or at least an efficient self-organizing system. In real life, however, we find that a more visible hand is required to organize people, to inventory people's opinions, mobilize support and finance, and so on. Especially when it comes to the screening and matching part of the process, an organization is required that is respected and supported by the participating actors. The organization has to have sufficient respect to make sure that (even the unpopular) recommendations or decisions are supported by every participant. Depending on the bottlenecks of the area to be developed, a number of options are available, ranging from a government agency to a local producers' co-operative, a consortium of local banks and chamber of commerce, and so forth. What is possible and desirable depends on the nature of the local networks – both material and immaterial – and mental maps, that is, the threats *and* the opportunities for the area concerned. The worst recommendation here would be to give a universal solution that would apply for every situation. We only want to emphasize that *some* type of respected institution, preferably including the support of local banks, has to play a role here.

3.9 The case of the farm banquets[4]

The North-Eastern part of the Netherlands named the *Veenkoloniën* can be characterized as an area with important but declining agriculture. Large farms with wonderful manor type farmhouses dominate the area. However, declining revenues from agriculture have forced the farmers to look for other opportunities. A small and active group of farmers in the village of Borgercompagnie decided to set up the *Compagniester Banquet*. They listed all activities planned, and local suppliers available, and checked on possible missing activities. They wanted to arrange for tourists to come to the North by train and arrive at one of the stations in the area. There, they would pick up a bicycle – typically Dutch – and cycle to the first farm

---

[4]   Based on interviews and on van Broekhuizen et al. (1997), van der Ploeg et al. (2002), and on <www.banket.org>. There are many other examples of reverse network engineering in present day rural development in Europe; see, for instance, van der Ploeg et al. (2002).

where they would have a guided tour around the farm, followed by the first course of a dinner. After that, they would continue to the second farm at cycling distance for the second course, and so on. The standard trip included a visit to the local church, a botanical garden, and a local museum. The concept produced can be described as leisure, culture and pleasant scenery, while being actively involved. In a time of wealthy but active 45+ tourists, the organizers thought it would be a market. In case of rain – quite possible in Holland – they had a horse-drawn carriage as back-up. In addition, they had a van for bicycle repairs and a support car for possible dropouts. The bicycles, carriage, and repair van were supplied by local suppliers. And if the guests did not like cycling, the carriage could be reserved as well. The meals were prepared by the participating farmers, with some professional help, with a focus on local produce. Further arrangements could be made, possibly combined with a stay in a luxury hotel annex golf resort in the North. The farmers, usually well-educated and able to speak various languages, did the guided tours. To avoid the costs and power of travel agents, the marketing and sales were pioneered using the Internet (www.banket.org). This was quite a novelty in the years when the experiment started (the first banquet was held in April 1995) – around the beginning of the Internet – when many private persons still did not have a PC at home and using the Internet was not common practice. Today, the number of people involved in the organizations amounts to about 160. The concept was a big success; initially many tourists from abroad, especially Americans, came and participated, booking through the Internet. Gradually, the emphasis moved towards groups: family days, personnel trips, and rallies. By listing all products and activities involved, organizing all activities – mostly performed by selected third parties – and carefully coordinating everything, the group managed to make the idea work. The banquet was the start of a successful trajectory of rural development in the 1990s: the organizers also developed an art-museum route, set up four mobile selling points of local produce, an Internet shop for food products, and two agro-food shops in the town of Groningen.

3.10 Summary and conclusions

The method described was developed for tourism in peripheral areas. The example of the farm banquets in the North-East of the Netherlands shows many of the dimensions we mentioned. It shows the separation of the area with the *markets* for products – the origin areas of both the American tourists and also the many tourists from other parts of the Netherlands – and the actual production/consumption area. It also shows the role of telecommunication to bridge distances. Finally, its shows the importance of combining the top-down approach – focussing on the type of network and the type of activities required – and the bottom-up approach – focussing on the question of what is suitable, feasible, and acceptable to the actors inside the production area.

The same methodology can be used as part of an industrial development policy or even by an individual company trying to set up a plant somewhere else. In that case, the actual existing network in an area can be matched with the required supplier network or with the profiles of the actors required. A special case would be a turnkey project where the availability of (local) suppliers has to be mapped or traced or where even a network may have to be developed 'out of nothing'.

To summarize the entire process described:

Step 1
1. Derive concepts from lifestyles and consumer patterns.
2. Derive the products that form the components or ingredients making up or associated with the concepts found.
3. Draw up an inventory of activities required to produce those products.

Step 2
1. Draw up the filière in economic space as the total of all sub–filières.
2. Use actors at an aggregate level: 'types' of actors rather than actual companies.
3. Indicate possibilities for economies of scope.
4. Mark potential situations of conflict between activities and concepts.

Step 3
1. Add geo–space by taking the locational preferences, restrictions and requirements of activities into account.
2. Check on positive synergy caused by economies of scale and scope.
3. Maintain markers on activities and their actors because of potential conflicts with certain concepts.

Step 4
1. Apply the result of Step 3 for an actual area and actual concepts.
2. Constantly check for suitability, feasibility and acceptability.
3. Explicitly screen the results and match with visions, objectives and perceptions.
4. Find out what actors want and distil a shared strategy.
5. Actors present in the area are named by their company name.
6. Missing actors are named by their category name.
7. Include scope effects.
8. Check on conflict situations, and work out options for solutions.
9. List restricting effects caused by unacceptable activities present in the area.

Results
1. Blue-print for the development of an area in terms of activities and their interrelationship and interdependencies.
2. Iinsight into the acceptability, suitability and feasibility of the plan.
3. Policies to create certain actors presently missing in the area.

4. Policies to encourage certain actors to relocate activities from the area that are conflicting with a selected concept, while maintaining the area as a market for the specific actor.
5. Policies to discourage certain actors to continue activities in the area since they are a restricting element and stimulate them to relocate to areas elsewhere where they do not upset the tourist 'atmosphere' associated with certain concepts.
6. Policies to stimulate transport and telematics for those activities that can, or have to be, performed elsewhere either because of locational requirements or because of conflicts between concepts being supplied.

Prerequisite
A generally respected institution that has roots in, and involvement with, the local actors has to play a pivotal role in organizing and mobilizing opinions and resources, while taking proper action to suggest strategies to deal effectively with threats in order to seize opportunities.

References

Axelsson, B. and G. Easton (eds) (1992), *Industrial Networks*, Routledge, London.
Borgstein, M.H., R.P.M De Graaff, J.H.A. Hillebrand, J.F. Scherpenzeel, F.J. Sijtsma and D. Strijker (1997), *Ketens en plattelandsontwikkeling*, (Chains in Rural Development), NRLO-report, The Hague.
Broekhuizen, R. van, L. Klep, H. Oostindië and J.D. van der Ploeg (1997), *Renewing the Countryside*, Misset Publ., Doetinchem, pp. 83–5.
Button, K., H. Priemus and P. Nijkamp (eds) (1998), *Transport Networks in Europe*, (NECTAR publication), Edward Elgar, Cheltenham, UK.
Camagni, R. (ed.) (1991), *Innovation Networks: Spatial Perspectives*, Belhaven, London.
Checkland, P. and J. Scholes (1990), *Soft Systems Methodology in Action*, John Wiley & Sons, Chichester.
Flood, R.L. and M.C. Jackson (1991), *Creative Problem Solving: Total Systems Intervention*, John Wiley & Sons, Chichester.
Ford, D.I. (1978), 'Stability Factors in Industrial Marketing Channels', *Industrial Marketing Management*, 7: 410–422.
Johnson, G. and K. Scholes (1989), *Exploring Corporate Strategy*, Prentice Hall, Englewood Cliffs.
Jurna, B. and D.J.F. Kamann (2003), 'Supply Base Genesis in a Greenfield Situation', Paper presented at the *IPSERA Conference*, Budapest, 14-16 April.
Kamann, D.J.F. (1988, 2003), *Externe Organisatie* (Industrial Organization), Charlotte Heymanns Publishers, Groningen.
Kamann, D.J.F. (1997), 'Policies for Dynamic Innovative Networks in Innovative Milieux', in R. Ratti, A. Bramanti and R. Gordon (eds), *The Dynamics of Innovative Regions: The GREMI Approach*, Ashgate, Aldershot, pp. 367-391.

Kamann, D.J.F. and D. Strijker (1991), 'The Network Approach: Concepts and Applications', in R. Camagni (ed.), *Innovative Networks: Spatial Perspectives*, Belhaven Press, London, pp. 145–73.

Kotler, P. (1991), *Marketing Management,* 7th edn, Prentice Hall, Englewood Cliffs.

Pfeffer J. and G.R. Salancik (1978), *The External Control of Organizations*, Harper & Row, New York.

Ploeg, J.D. van der, A. Long and J. Banks (eds) (2002), *Living Countrysides. Rural Development Processes in Europe: The State of the Art*, Elsevier Bedrijfsinformatie, Doetinchem.

Porter, M.E. (1985), *Competitive Advantage*, Collier–Macmillan, London.

Porter, M.E. (1990), *The Competitive Advantage of Nations*, The Free Press, New York.

Tan, K.C., V.R. Kannan and R.B. Handfield (1998), 'Supply Chain Management: Supplier Performance and Firm Performance', *International Journal of Purchasing and Materials Management*, August: 2–9.

Wasserman, S. and K. Faust (1994), *Social Network Analysis*, Cambridge University Press, Cambridge.

Wernerfelt, B. (1984), 'A Resource-based View of the Firm', *Strategic Management Journal*, 5: 171-180.

Williamson, O.E. (1975), *Markets and Hierarchies: Analysis and Antitrust Implications*, Free Press, New York.

Chapter 4

The Potential of Virtual Organizations in Local Tourist Development

Anastasia Stratigea, Maria Giaoutzi and Peter Nijkamp

4.1 Introduction

The dynamic role of tourism in local economic development is greatly recognized, since it consists of a structural element of modern societies and is largely influenced by the radical changes occurring in the field of Information and Communication Technologies (ICTs).

The introduction of ICTs and their role in providing access to remote peripheral areas has enhanced the opportunities for these areas to enter the tourist market. In this context, the new technological regime brings to the fore new tourist resorts, which are competing on the basis of the variety of services they provide. The ICT developments have also revolutionized the entire tourist industry by generating new business models, by changing the structure of tourist distribution networks as well as by re-engineering all related processes.

In such a continuously expanding tourist market, small firms are striving for survival with the support of large tour operators, travel agents, and large hotels, which have a comparative advantage due to their better access to know-how, technological infrastructure, resources and market penetration. Consequently, nowadays the role of entrepreneurship is attracting much interest and gaining in importance.

Entrepreneurship has become a fashionable topic in modern industrial organization literature. Technological innovation, globalization trends, open competition and the emergence of new markets have acted as driving forces for the increasing interest in our entrepreneurial economy (Cooke, 2002; Dicken, 1998). The tourist sector forms an interesting and convincing illustration of the wide-ranging impact of new forms of business in the global competitive market. The market dominance of SME firms in this sector has vanished to the benefit of large multinational corporate business firms that deploy regional power and logistic networks to get a strong foothold in many tourist destinations. The modern ICT sector plays a critical role in this context, as it stimulates flexible specialization and virtual business organizations. Such virtual organizations have assumed a prominent place in the modern tourist industry.

The tourist sector is traditionally dominated by many small-scale business firms. In order to cope with the increasing market competition, *small tourist firms* are forced to focus on the adoption and use of those technological developments will allow them to acquire the critical mass of know-how and technology that supports and strengthens their position in the tourist market.

For this purpose new organizational schemes supported by ICTs, such as *virtual organizations*, have become strong policy tools. These are organizations characterized by their product-market strategy, network structure, information systems and business communication patterns, as a response to increasing competition and a need for efficient use of resources.

The focus of this chapter will be to explore the potential role of virtual organizations on local tourism development, with emphasis on policy aspects in remote peripheral areas. Section 4.2 presents the tourist sector in perspective and attempts to outline relationships among small tourist firms and large tour operators. Section 4.3 elaborates on virtual structures and their potential contribution to small firms in the context of local tourist development. Section 4.4 explores policy directions appropriate for the promotion of local tourist assets. Finally, in Section 4.5 a number of conclusions are drawn.

4.2 The tourist sector in perspective

During the last decades there has been a steadily growing trend in modern societies towards *leisure time activities*, which have been to a large extent the result of:

- a steadily increasing *income* available for recreation activities;
- an increase in the *time* available for such purposes;
- an increasing *mobility* of people due to the shrinkage of distance resulting from technological developments in the transport sector;
- the *expansion* of the transport system towards new destinations;
- the change of *behavioural patterns* of people related to travelling, due to increasing internationalization in the information era;
- the advances of the ICT sector.

The tourist sector in this context has become one of the major sectors in most local economies, mainly due to its increasing share in income distribution but also due to the opportunities it provides for upgrading local development potential. Tourist development has become one of the main policy paths towards regional development, especially for peripheral areas with resources for tourist development, such as tourist infrastructure, quality of the environment, climate, transport infrastructure, etc.

Currently, the *positive impacts* of tourism on regional development are greatly recognized. Moreover, as the most important export industry in the world, the tourist sector often represents one of the main sources of income.

At the same time, one should be aware of the *negative impacts* in the development of tourism, which are related to social aspects, cultural aspects, and the environmental aspects of host areas. Since the tourist product is a complex product composed mainly of the above factors, it is clear that any development in tourism that shows no respect for these product ingredients will be only short-term.

As a result of the above debate, the concept of sustainability of tourist development has come to the fore. *Sustainable tourist development*, as a new popular concept, refers to tourist growth which at the same time prevents degradation of the environment, a fact which may have major consequences for the future quality of life (Nijkamp, 1997).

The above-mentioned problems regarding the positive and negative impacts of tourist development have been embodied in the approaches of various researchers when studying the effects of the tourist sector at a regional level. Strict sectoral economic approaches of the local economy have thus been replaced by a more general view, where economic and non-economic effects of the various sectors are brought together and can be classified under the following three distinct categories (Janssen and Kiers, 1990):

1. the *efficiency* aspect, referring to the economic side of the local society, e.g. allocation process, markets, financial systems, production of goods and services, etc.;
2. the *equity* aspect, associated with the social side, e.g. labour, income, employment, social structure, etc.;
3. the *conservation* aspect, related to the conservation of cultural/historical values as well as environmental issues.

It is evident that efficiency, equity and conservation impacts of tourism development on the local structure are closely related. In the triangle model in Figure 4.1 (Janssen and Kiers, 1990), these interactions are clearly presented. The main interaction between efficiency and equity mainly refers to the demand and supply of labour and of goods and services. Interactions between efficiency and conservation are related to the demand for natural resources and land use as well as to the supply of waste. Finally, the main interactions between conservation and equity refer to requirements for well-being and waste.

Although the tourist product provides the basis for the successful tourist development of a region, it is not sufficient. *Organizational structures* of the tourist firms and their potential to promote their 'product' through *innovative marketing approaches* constitute a crucial factor for their development, and here the role of ICTs is greatly acknowledged. The introduction of various *applications* related to the *tourist sector* opens new horizons for the development of new tourist services of either existing or newly emerging tourist resorts in *peripheral areas*, which will strengthen their position in the tourist market. Such applications may focus on:

- the *promotion of tourist destinations* based on the advertisement of the tourist product through multimedia applications;
- the *interactive communication* between the interested parts (tourist destination and the tourist);
- *on-line transactions* between the tourist destination and the tourist, such as booking, payment, etc.;
- *teleworking applications* that provide the opportunity for tourists to combine work with vacations and thus eventually enlarge the duration of leisure time;
- *telemedicine applications* that enable aged people to prolong their stay away from home;
- *transport telematics* that facilitate more efficient management of the tourist flows etc.

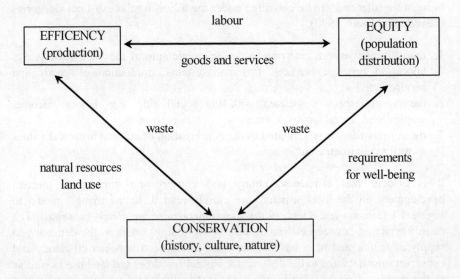

Figure 4.1   Triangle model of tourist development
Source: Janssen and Kiers, 1990.

The enormous developments of the tourist sector as a result of these applications have considerably emphasized the tourist *supply side*. Certain aspects of these consequences are listed below (Janssen and Kiers, 1990):

- The *tourist destinations* are very much influenced by various factors, such as the role of tour operators, while new destinations are coming on the scene.
- The organization of the *supply market* has been further developed and transformed into a very well-organized, highly networked market with strong competition, especially for tourist SMEs.
- The modern *information* and *communication technologies* represent the core of promotion and marketing programmes within the tourist market. Their role is also essential in terms of accommodating the communication costs involved. This fact favours large tourist companies with access to such technologies.
- Demand is growing for well-organized *modern*, *high quality* and rather *diversified tourist activities*.
- The need for well-educated and rather skilled *labour* involved in the supply of tourist activities becomes a necessary and sufficient condition for surviving in the tourist market. This remains one of the main disadvantages of peripheral areas, which do not have such labour potential.
- Demand is growing rapidly for more *active types of tourism* closely related to nature and socio-cultural activities.
- An increasing demand for *cultural tourism* has developed over recent years.

Today, the *tourist sector* can be considered as a very dynamic but sensitive sector, where *control relationships* are rather prominent. This is especially true in the context of peripheral or less developed areas, where tourism follows the patterns of the international division of labour between developed and less developed regions of the world. This division is greatly increasing the *dependence* of these countries, since tourist demand has been largely based upon the cyclic fluctuations of the economies of developed countries (Komilis, 1982).

Control relationships become even stronger where the tourist market has been controlled by *large foreign travel mechanisms*, such as tour operators, large hotels, etc., with a direct impact on tourist sector SMEs. This is rather crucial, especially for peripheral areas, since:

- it deprives them of an independent development trajectory, which would have positive impacts on local economies;
- it involves certain risks for peripheral areas as destination regions, since at different times various exogenous factors including political, financial strategies and profit of large firms may determine their tourist flows;
- it deprives them of their right to decide on the type of 'tourist product' they want to promote (mass versus individual), and consequently the type of tourists/visitors they are willing to host. This may result in rapid and uncontrolled tourist development, which will not be in harmony with their physical assets, social reality, culture and the environment, etc., and thus will not act as a factor of stimulation for their local identity (Boissevain, 1985).

The development of local tourist SMEs provides the vehicle via which peripheral areas may overcome such bottlenecks, since they assure local participation in the production and management of the tourist product, thus determining to a great extent the quality and intensity of socio-economic impacts (Kokkosis and Tourlioti, 1998).

The efficiency of local tourist SMEs in relation to their competitive advantage within the international tourist market is to a large extent dependent upon their *access* to ICTs, which is conditioned by their capacity to afford the *costs* of using the networks and their *expertise* in applying network facilities. Small tourist sector SMEs do not usually meet the above requirements and thus are forced to gain access to the tourist market through large tour operators at very high costs.

New organizational schemes adopted by local tourist firms together with appropriate *policies* may reverse the existing trends. In this respect, small tourist firms face a great challenge when supported by ICTs both as organizational tools and as promotion strategies. These have the potential to redefine the roles of the different tourist actors and overturn the domination of large travel agents, tour operators, etc.

The next two sections will focus on new organizational schemes and more specifically *Virtual Organizations* (VOs) coupled with specific *policies* to support SMEs in the tourist sector.

4.3 Virtual organizations: A new organizational perspective for the tourist sector

The term 'virtual organization' refers to an actual organization which, like traditional corporations, aims to supply goods or services by means of cooperation between independent enterprises. Along the chain formed by the different enterprises participating in this organization, partners bring in their critical core competencies (skills, capabilities, resources, know-how, etc.). The output is the integration of an ideal combination of best-of-class (complementary) core competencies (Zimmermann, 1997) into a 'best-of-everything' organization for the specific purpose (Byrne et al., 1993).

The *virtual corporation* concept is evolving as a specific example of a flexible networked organization. It actually represents a *new strategy*, supported by ICTs, which aims to realize short-term flexible cooperation schemes both horizontally and vertically structured, where the culture of control is replaced by the culture of information and knowledge sharing (Jagers et al., 1998). Their strength is in *integrating* different parts/enterprises in schemes with a high degree of horizontal structuring, i.e. flat hierarchies, low formalism, and strong team orientation. In this sense they are multicentric, or even 'centreless'.

The issues of *short response* to the market demand and more *flexible adaptation* to the customer needs – crucial aspects for the *tourist sector* – have led to the necessity of developing such flexible organizational schemes, which would fulfil the above requirements, in particular for small and medium-sized enterprises. Virtualization structures in general offer organizations several types of *benefits* such as (Skyrme, 1998):

- they can source intellectual *resources* globally;
- they can gain *flexibility* through dynamic structures and contractual agreements;
- they can tackle *projects* or *problems* which might otherwise have been beyond their capabilities;
- they can reach *global markets* without a local presence;
- they can significantly *reduce costs* over conventional ways of working, which are for the benefit of both large and small/medium-sized enterprises.

These can be of great importance when placed in the context of the tourist SME sector.

In the following passage, a classification of virtual organizations (VOs) will be presented in order to explore types of VOs, which better apply to the tourist sector. VOs can be classified as *static* and *dynamic* (Franke, 1998), and can be further classified as follows (Figure 4.2) (Campell, 1997):

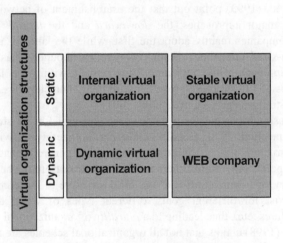

Figure 4.2 Dynamic and static virtual organization structures

An *internal VO* is a type of virtual corporation applied *within one organization*, which aims at creating flexible organizational structures. In this case, the virtual enterprise consists of several business units, built by autonomous groups and teams.

A *stable VO* is another type of organization, structured on the basis of *inter-organizational* cooperation among different companies. It is based on the concept of contracting non-core competencies from a main organization – the *core partner*. Several firms provide input or distribute the output of the core organization. In this context, a cooperation scheme emerges, in which several committed suppliers are closely related to the main (core) organization, which leads the whole scheme.

Other VOs, which have been characterized by their *dynamic nature*, are leading to *dynamic VO schemes*. The 'dynamic nature' implies that the entire set up of a VO may change in response to the marketplace. In this sense, VOs of this type are *temporary* in regard to their ability to react quickly in terms of membership, structure, objectives, etc. *Vague/fluid boundaries* and *opportunism*, as well as *equity of partners* and *shared leadership*, are the main characteristics of a dynamic VO. Experience shows that this type of organization can only be established in the case of small companies. It should be mentioned that most definitions of VO encountered in the literature refer to dynamic and WEB-company VOs.

The potential offered by the Internet has provided the grounds for the development of the *WEB-company*. This company consists of a *temporary network* of a *dynamic nature* formed among specialized organizations based on the use of Internet. The aim of such a company is to offer all sorts of products and services on a global scale by *using the Internet*. The WEB-company has the same characteristics as the dynamic VO, is characterized by vague/fluid boundaries, and consists of a well-suited scheme for SMEs. The difference from the dynamic type is that the WEB-company is fully based on *information technology*.

Ching et al. (1993) point out that the establishment of network organizations follows two major approaches: the *downward* and the *lateral*. Large vertically integrated companies mainly adopt the first, while the latter is adopted by small and medium-sized companies. Static VOs can be thought of as falling into the downward approach while, dynamic VOs remain within the lateral approach. Various types of VO structures, together with their characteristics, are presented in Figure 4.3 (Stratigea and Giaoutzi, 2000, 2001).

In any case, although VOs provide great potential for the strengthening of a company's position in a globally changing competitive environment and a prominent structure for the future, they are not a general panacea. As Drucker (1997, p. 4) points out: 'Researchers are moving away from the belief that it has to be one theory of organization and one ideal structure'. Different organizational structures exist for different goals, different types of work, different people, different cultures, etc., thus leading to a *plurality* of organizational forms.

Dembski (1998) claims that not all organizational schemes are relevant to every kind of cooperation. He therefore attempts to classify the classical and new organizational schemes according to whether they: 1) are *intra* or *inter-organizational* schemes; 2) concentrate on *economies of scale* or *scope*; 3) result in the *stability* or *flexibility* of the system, etc. (Figure 4.4). He points out that the VO is a perfect solution in a business environment that requires both broad strategic and organizational focus. Also VOs thrive on the economies of scope, which requires intense inter-organizational cooperation and a high level of flexibility.

		Type of virtual organization	Management scheme	Type of companies involved	Time horizon	Approach
Virtual Organizational Structures	Intra-organizational	Internal V.O.	hierarchy (decentralized to autonomous teams)	large corporation	long/permanent (static nature)	downward
		Stable V.O.	hierarchy (core partner)	large corporations	long/permanent (static nature)	downward
	Inter-organizational	Dynamic V.O.	market (shared leadership)	SMEs globally dispersed	short/temporary (dynamic nature)	lateral
		WEB company	market (shared leadership)	SMEs globally dispersed	short/temporary (dynamic nature)	lateral

Figure 4.3 Types of virtual organization structures and their characteristics

As Zimmermann (1997) points out, the virtual enterprise is a crowning of organization development in a turbulent, high-speed, competitive environment, imposing extremely high requirements on adaptability, flexibility and reconfigurability. However, not all companies can thrive in such an environment. On the contrary, there are industries working in a stable environment, where scale economies are present, and the old-familiar hierarchical organization remains unchallenged. Figure 4.4 reflects the various business-specific organizational schemes.

Sieber (1996) advocates the idea of business-specific organizational schemes by arguing that VOs are not the appropriate form of organization in a stable market environment where there are no technological upheavals. He classifies the various forms of organization according to both the *variability* and the *specificity* of the tasks they handle (Figure 4.5).

Tourism constitutes a *complex economic product* based not only on the tourist sector but also on other economic sectors as well. In such a context, tourist managers transport people to different places to enjoy natural features, cultural attractions, economic activities and life-styles of people worldwide (Tsolakidis, 1998).

Due to the multisectoral and international nature of the sector, tourist operators cannot control the full chain of its production, since such an effort involves coordination of different sectors, spatial levels and tourist products. Instead, they need to seek for more effective new strategies in order to plan, manage, promote and sell the tourist product at an international level. More specifically, it seems that tourism requires a practical and sustainable opening up to the world of giving and receiving information, of communicating and making agreements, of coordinating and exchanging ideas and thoughts, of incoming and outgoing money flows

(Tsolakidis, 1998). VOs, supported by the advances of ICT, may play an important role, especially by reinforcing the position of SMEs in the tourist market.

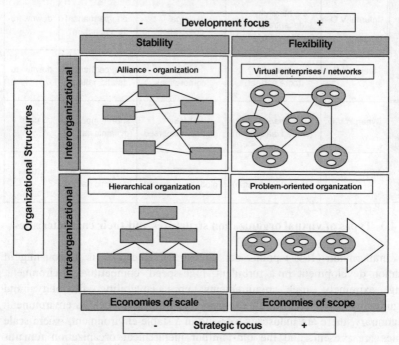

Figure 4.4 Business-specific organizational structures
Source: Dembski, 1998.

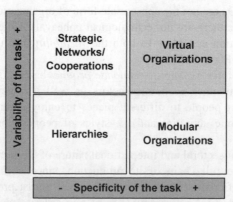

Figure 4.5 Forms of coordination and task structure
Source: Piecot et al., 1996; Griese and Sieber, 1996.

Such organizations in the tourist sector may be seen from two different angles:

1. the *intra-organizational angle* related to virtual structures among businesses in various sectors which have tourist companies at the top of their chain;
2. the *inter-organizational angle* related to different tourist firms which cooperate in order to produce a more rich and diversified tourist product.

The *intra-organizational angle* relates to virtual structures established among firms of different sectors, having tourist companies as the *core partners* in the chain. It constitutes an *inter-organization cooperation* among companies of different sectors, such as transport companies, hotels, travel and tourism agencies, restaurants, handicraft companies, museums, galleries, public institutions, entertainment, various services, etc., where tourist companies lead the whole organizational scheme. This may be both of a *stable form*; for example, in the case of remote and peripheral areas characterized by production factors of limited size and diversity, as well as lack of economies of scale, and a *dynamic* one in more centrally-located areas. It follows a *lateral approach*, which applies among SMEs of the various sectors, located mainly at a *local/regional spatial scale*. These may contribute to the effort of local/regional areas to keep *control over their tourist product*.

The *inter-organizational angle* relates to virtual structures formed among tourist firms. Such an organization may refer to both the *regional* and *inter-regional spatial level*. It may be of a *stable* form, having a *core partner* leading the whole chain, for example, several tourist companies at a regional level. It may also be of a *dynamic* form established, for example, among tourist companies of various regions, e.g. the Mediterranean regions, in order to promote a more rich and diversified inter-regional tourist product. It may adopt a *lateral approach*, which applies among *SMEs* of the tourist sector. It may also apply among VOs of local interest (intra-organizational schemes), thus forming a *second level of VO* (inter-organizational schemes) (Appel et al., 1998: see Figure 4.6).

4.4 Policies encouraging the development of local tourist SMEs

Tourism is one of the most significant components of leisure society, nowadays being a mass phenomenon with increasing trends towards an individualistic tourist attitude, differentiated according to age, income, culture, leisure-orientation, etc. Those tourist resorts which offer a wide variety of services to individually oriented clients may be expected to become winners in this strong international competitive game.

The current individualization trend in tourist behaviour, supported by ICTs, gives remote areas a unique opportunity to develop tailor-made services focussed on the cultural and environmental needs of the modern tourist.

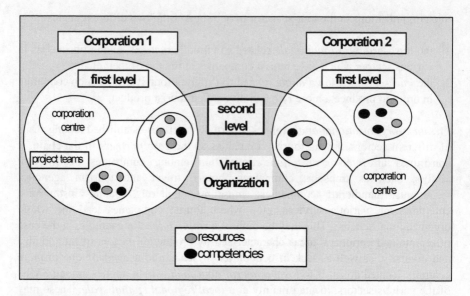

Figure 4.6   First- and second-level virtual organizations
Source: Appel et al. 1998.

VOs established in such a context may result in multilevel regional cooperation, mutual satisfaction but also more attractive attributes: richer and more comprehensive tourist offers, bigger market segments and, finally, better financial results. Within such schemes, tourist enterprises can provide access to cheap, exclusive and first-class advertisements for an enormous number of potential tourists (Tuntev, 1998).

Sets of policies encouraging likely developments towards VO structures will be presented in the following section, which may include (Giaoutzi and Stratigea, 1997):

- *Multimedia applications* bringing tourist and cultural information to worldwide markets in the form of images, sound and data at far lower effort and cost. This increasing efficiency in tourist operation may also act as further stimulus for enhancing but also reorienting tourist flows.
- *Electronic Commerce applications* allowing direct bookings and payment transactions between tourists and host places.
- *Teleworking applications* enabling tourists to combine work and pleasure in remote localities and extend the duration of their vacations.
- *Telemedicine applications*, which may encourage aged people or those with weak health to enjoy their holidays away from home.
- *Telematic applications in transport*, which may facilitate the smooth flow of tourists and upgrade the transport logistics.

- *Telematic applications in libraries*, which may enhance the flows of cultural tourists in remote areas.

The above applications highly contribute to the *efficiency* and *demand management* of the tourist sector while at the same time they diminish the importance of distance for remote regions, such as islands, which may have far-reaching consequences for their economic, cultural and social life.

Policies encouraging such applications, which will support the development of VOs, focus on the following aspects of policy intervention.

4.4.1 Infrastructure provision

The beneficial role of ICT on the development perspectives of the various regions has been mainly based on the vital role of *information exchange* in the context of information economy.

ICT *infrastructure* development policies should aim at the deployment of national information infrastructure on the following aspects:

- the establishment of *telecommunication infrastructure*;
- the establishment of *content-specific strategic information systems*;
- the establishment of *transport infrastructure* via alternative transport modes supported by telematic applications.

The above aspects are necessary for widespread access to communication and use of information services, which will enable the development of virtual applications not only in the tourist sector but in any branch of social and economic activity, especially in remote peripheral areas.

In order to achieve the goals of providing *universal access* of regions to ICT infrastructure, several policies can be adopted which aim at providing rural, remote and disadvantaged regions (that is, the less profitable regions versus urban regions) access to reliable and high-quality telecommunication infrastructures.

Socio-economic and demographic diversity implies also region-specific goals related to policies for *effective* and *equitable infrastructure provision*, especially in the light of the liberalization of the telecommunications market. These may focus on the following schemes (Hudson, 1997):

- Provision of *incentives*, through competition or concession to telecommunications operators.
- Policies to foster investments and competition in the telecommunications industry in order to assure continuous innovation and aggressive pricing of services.
- Requirements of licences or franchises, like the *universal service obligation* (USO) condition as a prerequisite for licence provision to operators, applied by many countries, or the *carrier of last resort model* via which the dominant

carrier, entitled to a subsidy, has the obligation to provide the service if no other carrier will.

- Various forms of *subsidization* of services in areas which are thought of as less profitable to serve. Subsidies may apply in both *high cost areas* (isolated areas, areas with very low population, etc.) and *disadvantaged areas* or to *customers* related to areas or groups which cannot afford the costs of gaining access to and using such services. Some countries are applying the *route-averaging model*, which requires all rates to be averaged so that every customer is paying uniform distance charges regardless of the location.

4.4.2 Cost of use of ICT

A very important aspect in the application of ICT in remote areas concerns the capital and operating costs, that is, the costs of getting properly equipped (the necessary hardware and software) and the use of such technologies. These costs may include pricing of installation, monthly service, local and international communication costs, etc. This applies especially to new SMEs and, to a great extent, it influences the entrepreneur's decision to join the network. Policies to overcome this problem could include:

- incentives related to tax release, partial subsidization, etc. for being *properly equipped*;
- incentives which encourage entrepreneurs to *join the network*, e.g. subsidizing the costs of use for certain periods;
- *provision of information* on the firm's economic benefits involved as a result of such a decision, which may prove beneficial for the firm in economic terms by increasing its market penetration.

In all cases, it should be noted that the costs of using such communication technologies are constantly declining and are becoming increasingly distance-independent as in the case of satellite transmission. In this context, rural and remote users will not be penalized, since very soon billing will be related to time or unit of information transferred and not to distance. Nevertheless, there is always the problem of continuously upgrading the existing infrastructure in order to meet the continuously upgrading technologies; this involves a heavy economic commitment for the firm or individual.

4.4.3 Education

Technological advances may also support the *capacity* of each region to build, operate, manage, and serve the technologies involved. Not all regions exhibit the same level of expertise in accomplishing these functions, which may be considered as an important comparative region disadvantage. Policies that aim to increase the capacity of each region to join the network may include:

- *upgrading of the capacity* of the *workforce* to adequately develop, maintain and provide the value-added products and services required by the information society;
- supporting *life-long learning processes* through tele-education procedures in connection with universities, research centres, etc.;
- supporting *links* with local research institutions, universities, etc., which may serve as information and education centres and as region-specific hubs operating at low or subsidized cost for local users;
- providing *education* and *training* focussing on *specific themes* like use of networks, tourist policies, tourist catering professions, resource management, small and medium-sized entrepreneurship, etc.

4.4.4 Create more efficient local/regional government structures

Public administration structures may play a very important role in compromising different views of tourist development and promoting specific tourist policies. Efficient and fully-equipped local/regional government structures will be more effective in collecting, processing and diffusing information to both public and private domains related to (Tsolakidis, 1998):

- information about the specific community/municipality;
- information about tourist attractions of a certain region;
- general information about the whole region;
- information about the local tourist product by type of tourism;
- information about various tourist events, and scientific and cultural attractions;
- information about local tourist infrastructure (hotels, travel agencies, transport facilities, etc.);
- information related to the formation of VOs with other tourist destinations/ agencies;
- information related to the demand side of tourism (origins of tourists).

In the process of handling and successfully disseminating information by public authorities, being both producer and consumer of knowledge and information, some very important policy aspects should be included. These are:

- modernization of *equipment* at a local, regional and national level;
- modernization and improvement of the *quality of services* offered by the public at a local, regional and national level;
- provision of *public services through the network*, which will remove obstacles imposed by distance and transport inefficiency and ensure *social inclusion* (for example, islands, and isolated or peripheral regions) and at the same time increase the efficiency of tourist firms located in the region;

- making *better use of information* on tourist issues by creating *public community tourist information centres*, which will provide tourists with valuable information on the various opportunities appearing within the region;
- policies building *services* around *tourists' needs*: for example, creating teleworking centres with ICT facilities, so that tourists can communicate with their countries and businesses while on vacation.

4.4.5 Creation of an information-friendly environment

Another very important policy aspect is related to the creation of a proper *physical setting* in which information is produced and diffused. Such policies are based upon the creation of an *information-friendly environment*, which supports availability, diversity and low-cost information services and products. These are very important aspects for the formation of VOs in all sectors. Policies supporting such purposes can include (Talero and Gaudette, 1996):

- introducing coherent *telecommunications reform* and the promotion of information dispersion policies;
- introducing *legislation policies* promoting laws which protect local investment, intellectual property rights and security;
- creating open and well-regulated *information* and *communication markets*;
- launching effective *regulatory* and *standard-setting institutions*;
- encouraging close *cooperation* of tourist firms with local universities, research institutions, etc.;
- providing incentives for the location of ICT companies familiar with relative ICT applications, e.g. tourist applications that support any business effort towards the adoption of such applications;
- favouring the education of the local society in the importance of ICTs.

4.4.6 Creation of a public brokerage platform to encourage virtual perspectives of tourist initiatives

The role of a public brokerage platform can be of great importance especially for the tourist development of disadvantaged or peripheral regions. In such a context, national tourist organizations may develop a *brokerage platform* contributing towards the *adoption* and *diffusion* of the *virtual perspectives in the tourist sector* by facilitating access to and use of ICT, and coordinating various efforts either locally or inter-regionally, through:

- providing *access to ICT* for SMEs and acting as a hub connected to multiple self-sufficient nodes, e.g. international tourist organizations, regional tourist SMEs, etc.;
- providing *access to content-specific strategic sector-wide information systems* related to tourism and other complementary activities (content provision);

- collecting and dispersing information on *opportunities* for SMEs to join *virtual schemes* and coordinating efforts for linking various actors/entrepreneurs willing to follow virtual perspectives (partnerships). This will assist local tourist firms to cooperate with other firms of related or complementary interests, either locally or inter-regionally;
- promoting strong and long-term institutional cooperation between national tourist institutions and international tourist associations;
- establishing the necessary *mechanisms* dedicated to furthering the use of the above technology and information by all tourist recipients.

More precisely, *access to ICT* for SMEs, who are not able to afford the costs of getting equipped and using ICT can be provided through various policies such as:

- installation of *community teleservice centres* at fully-equipped locations supporting tourist *SMEs* to evolve virtual relations in their activities, ranging from simple e-mail exchange and promotion through the network to participation in VO structures at a low or subsidized cost;
- establishment of costless (or partly subsidized) *web-site design services* for those firms interested in advertising their products/services, gaining access to the global marketplace, and searching for partners with which to form VOs;
- provision of technical support on technological aspects of various applications.

Another very important aspect of such a brokerage platform is providing *content-specific access to strategic sector-wide information systems.* Such information can be region-specific, and it may refer to education, health, public financial management, transportation, trade facilitation, university and science networks, local business characteristics, local and national statistics, key-economic sectors of various regions, products, electronic transactions, technological infrastructure, etc. This is valuable information both for locally-operating tourist SMEs and for tourist firms searching for new partners.

Finally this public-oriented brokerage platform is of great importance in collecting and dispersing information about *opportunities for joint virtual schemes* in the tourist sector as well as coordinating efforts for bridging various actors/entrepreneurs, who are willing to follow tourist virtual applications. It can monitor region-specific *databases* with *tourist-specific information* of firms seeking partners to create tourist VOs and also provide *consulting services* (e.g. legal aspects, technological aspects) with respect to new types of firms on organizational and development aspects (e.g. VOs).

It may also provide information about tourist successes and failures of remote areas elsewhere, as well as permanent monitoring and critical evaluation of the performance of each peripheral region in the tourist sector. This may be seen as a necessary action for a balanced regional development policy serving the interest of both local actors and the tourists.

Such a platform could be of great importance with respect to the development of specific tourist applications that provide assistance both for the design and the technological support of tourist SMEs.

Its role may focus on:

- The design of professional, regularly-updated, content-specific Web-sites capable of competing with other distractions on the Web.
- The design of various tourist applications as a means to stimulate and compel the *virtual tourist*. Potential tourists always prefer to get a sense of the place and the people they are interested to visit. Conventional media, such as printed material containing visual images and maps of the various places, have served this need up until very recently. This approach provides rather static and poor information in relation to intuitive and user-friendly tourism applications to support a more dynamic and empowered tourist service for clients.
- The design of specific *cultural tourism initiatives* and their presentation virtually through the appropriate virtual reality and multimedia software in order to take advantage of the worldwide trend towards cultural tourism. In this context they can approach clients with sophisticated needs, who are choosing their tourist destination on the basis of its tradition, art, cultural and historical heritage.
- The formation of networks with other organizations on the basis of a common promotion of their activities or cooperation on a cultural level.
- The design of *electronic commerce applications* for flight bookings in order to achieve a more equitable distribution of the financial benefits across home and host countries.
- The promotion of *corporate tourist presentations* of several countries or regions with common or complementary characteristics.

Finally, it is important that short-term local interests on an ad hoc basis do not dominate decision making in remote areas, which should be based on long-term strategic perspectives that attempt to reconcile the socio-economic and environmental interests of these areas.

4.5 Conclusions

Development of ICTs and their applications in the tourist sector have strongly influenced competition on a global basis. In such a competitive environment, large tourist corporations have the advantage over local tourist SMEs.

On the other hand, SMEs are facing new challenges in their dynamic entry into a continuously changing tourist market. These challenges are strongly supported by ICTs and may focus on both new organizational schemes (e.g. virtual organizations) and innovative promotion strategies.

Remote and peripheral areas are afforded great chances for promoting their tourist assets and development by the use of ICTs. In order to adhere to the goal of sustainable tourist development, these growth opportunities should be under the control of local potential in order to assure preservation and protection of the natural and cultural heritage of these areas. In this effort, local tourist SMEs are considered to be the vehicles for sustainable tourist development.

As power is heavily based upon the information and stock of knowledge available of regions as well as their access to ICTs, it is evident that equity aspects should be treated with great caution. Without a relevant policy framework, developments in the information economy are likely to exacerbate geographical divisions and strengthen regional disparities in the tourist sector as well. These policies should reinforce the role of ICT advances in the promotion of tourist assets and the development of remote and peripheral areas.

References

Appel, W., R. Behr and G. Hessen (1998), 'Towards a Theory of Virtual Organizations: A Description of their Formations and Figure', *Virtual-Organization.Net Newsletter*, Institute of Information Systems, Department of Information Management, University of Berne, 2(2): 15–36.

Boissevain, J. (1985), *Coping with Tourists*, Berghahn, Oxford/New York.

Byrne J., R. Brandt and O. Port (1993), 'The Virtual Corporation: The Company of the Future Will Be the Ultimate in Adaptability', *International Business Week*, 8 February: 36–40.

Campell, A. (1997), *Creating a Virtual Corporation and Managing the Distributed Workforce*, University of Paisley, Paisley, U.K.

Ching, C., C.W. Holsapple and A.B. Whinston (1993), 'Modelling Network Organizations: A Basis for Exploring Computer Support Coordination Possibilities', *Journal of Organizational Computing*, 3(3): 279–300.

Cooke, P. (2002), *Knowledge Economies*, Routledge, London.

Dembski, T. (1998), 'Future Present: The Concept of Virtual Organization Revisited: The Nature of Boundedness of Virtual Organizations', *Virtual-Organization.Net Newsletter*, Institute of Information Systems, Department of Information Management, University of Berne, 2(2): 37–58.

Dicken, P. (1998), *Global Shift*, Paul Chapman, London.

Drucker, P.F. (1997), 'Introduction: Towards the New Organization', in F. Hesselbein, M. Goldsmith and R. Beckhard (eds), *The Organization of the Future*, Jossey-Bass Publishers, San Francisco, pp. 1–5.

Franke, U. (1998), 'The Evolution from a Static Virtual Corporation to a Virtual Web: What Implications does this Evolution have on Supply Chain Management?', *Virtual-Organization.Net Newsletter*, Institute of Information Systems, Department of Information Management, University of Berne, 2(2): 59–65.

Giaoutzi, M. and A. Stratigea (1997), 'SMEs versus MNCs in Peripheral Regions: The Tourist Example', Paper presented at the *Conference on Tourism, Telecommunications and Regional Development*, Samos, 23-25 October.

Griese, J. and P. Sieber (1996), 'Die Virtuelle Fabrik: Ein Uberblick', *Computers and Security*, 15(6): 471–76.

Hudson, H. (1997), *Global Connections: International Telecommunications Infrastructure and Policy*, Wiley & Sons, New York.

Jagers, H.P.M., W. Jansen and G.C.A. Steenbakkers (1998), 'Wat zijn virtuele organisaties?: Op zoek naar Definities en Kenmerken', *Informatie en Informatiebeleid* (in Dutch), 1, Spring: 61–7.

Janssen, H. and M. Kiers (1990), '*Lesvos ... On its Way to the Future*', ERASMUS Project 1990, Free University of Amsterdam and University of the Aegean.

Kokkosis, H. and P. Tourlioti (1998), 'The Social and Cultural Impacts of Tourism to the Host Societies', Paper presented at the *First International Scientific Congress on Tourism and Culture for Sustainable Development*, Athens, 19-21 May.

Komilis, P. (1982), *Spatial Analysis of Tourism: Tourism Structure as a National and International Level*, Center of Planning and Economic Research, Athens, Greece.

Nijkamp, P. (1997), 'Tourism, Marketing and Telecommunication: A Road towards Regional Development', Paper presented at the *Conference on Tourism, Telecommunications and Regional Development*, Samos, Greece, 23-25 October.

Piecot, A., R. Reichwald and R. Wigand (1996), *Die Grenzenlose Unternehmung: Information, Organisation und Management*, Gabler, Wiesbaden.

Sieber, P. (1996), 'Virtuality as Strategic Approach for Small and Medium Sized IT Companies to Stay Competitive in a Global Market', in J. DeGross, S. Jarvenpaa and A. Srinivasan (eds), *Proceedings of the Seventeenth International Conference on Information Systems*, ICIS 1996, 16-18 December, Cleveland, Ohio.

Skyrme, D. (1998), 'The Realities of Virtuality', in P. Sieber and J. Griese (eds), *Organizational Virtualness*, Proceedings of the *VoNet – Workshop*, 27-28 April, Institute of Information Systems, Dept. of Information Management, University of Bern, pp. 25–34.

Stratigea, A. and M. Giaoutzi (2000), 'Teleworking and Virtual Organization in the Urban and Regional Context', in H. Bakis (ed.), *Build Space, New Technologies and Networks, NETCOM 2000*, 14(3-4): 331–58.

Stratigea, A. and M. Giaoutzi (2001), 'Virtual Perspectives as Policy Making Tools in Urban and Regional Planning', Paper presented at the *6th NECTAR Euroconference*, 16-18 May 2001, Helsinki.

Talero, E. and P. Gaudette (1995), *Harnessing Information for Development: A Proposal for a World Bank Group Strategy*, The World Bank <http://www.worldbank.org/html/fpd/telecoms/harnessing/>.

Tuntev, Z. (1998), 'Tourism and the Internet', Paper presented at the *First International Scientific Congress on Tourism and Culture for Sustainable Development*, 19-21 May, Athens.

Tsolakidis, C. (1998), 'Internet and Information Technology Means of Sustainability in Tourism', Paper presented at the *First International Scientific Congress on Tourism and Culture for Sustainable Development*, 19-21 May, Athens.

Zimmermann F.O. (1997), 'Structural and Managerial Aspects of Virtual Enterprises', in *Proceedings of the European Conference on Virtual Enterprises and Networked Solutions – New Perspectives on Management*, Communication and Information Technology, Paderborn, Germany.

Chapter 5

E-Travel Business: E-Marketplaces versus Tourism Product Suppliers

Dimitris Papakonstantinou

5.1 Introduction

Tourism enterprises need to re-examine profoundly their position in the continuously evolving chain of electronic commerce.

The tourism business has passed through several phases. During the pre-Internet period, before 1995, all tourism business was carried out manually, as it were, by means of the telephone. The dot.com phase, of 1999–2000, saw the creation of web-based tourism business. Now we have reached the period of e-Marketplaces. Tourism enterprises now have electronic points of sale and these are gradually evolving into web-based Virtual Travel Agencies.

The attention of tourism enterprises now rests mainly upon business-to-business (B2B) market models. This has come about because the Global Distribution Systems (GDSs) of airline companies, which represented the first serious attempt at the electronic distribution of tourism products, were, however, supported by out-of-date programming platforms, which suffered from a lack of integrated web technology. They therefore failed to satisfy the enormous operating needs of tourism enterprises. These needs have increased, along with the needs of the consumers, as technology develops at an ever-increasing pace.

In this context, the tourism industry needs to adopt more advantages provided by new technology as soon as possible, and tourism enterprises need to invest in the creation of their e-Marketplaces. At present, such e-Marketplaces are springing up everywhere, in a rather disorganized fashion, where strategic planning is almost absent. Boyer and Equilbey (1990) pertinently remark that: 'The art of management is to make the organization serve strategy.' At present, such sites simply serve to satisfy the entrepreneurial aspirations of their creators. For e-Marketplace to survive and thrive means, above all, complete control of the primary service offered by the supplier, at its source, so that it can be turned into a fully-integrated product.

The functioning of the e-Marketplace inevitably conflicts with the interests of suppliers of the tourism product. The problem is of vital importance for the future, since the ultimate shape that the travel e-commerce chain takes depends on how this problem is tackled.

Is there, then, a way out of the problem, while everyone is waiting to see what the major players in the tourism market will do? These players are the airlines, who have attempted to set the rules of the game via their GDSs. Initially, the airlines set these rules, solely at the retail level, although they have recently turned directly to the consumer.

Our aim here is to examine the problem, explore its dimensions and analyze the scenario where small B2B networks operate:

- In the first part, the problems involved in e-travel business are defined.
- In the second part, the factors responsible for shaping and influencing the e-travel business chain are traced.
- In the third part, a possible solution is offered, in the frame of the operation of small B2B closed networks, to the problem of the conflict between e-marketplaces and suppliers of the tourism product.

The viability of the tourism industry can be firmly assured, as long as the suppliers of the tourism product and the tourism enterprises agree to opt for the creation and support of e-Marketplaces for Business-to-Business e-Travel (that is, the subset B2B) by means of small closed private networks, consisting of Tourism Product Suppliers, Tourist Agencies and their Collaborators. This means that, after the new technology has been introduced and adopted, it will be possible to capture the field of business-to-consumer travel (the subset B2C), which, of course, is the ultimate goal. The suppliers will thus ensure their own viability.

5.2 An approach to e-travel business problems

Ensuring the viability of tourism is today a rather important issue at the international level. Any increase in revenues generated by tourism involves a number of issues such as: a redefinition of tourism resources; efficient management of these resources; and a rational process of making the product available, in terms of forecasting, promotion and sales. As a result, the aim of achieving an increase in tourist flows and revenues can be realized. There are, however, many weak links in the e-commerce chain: namely, among the institutions, public and private, that shape tourism policy, and the tourism enterprises that interpret their poor performance in terms of the heavy pressure of the competition in the international market.

In the business-to-business (B2B) chain, and the retail (business-to-consumer, B2C) trade, many private points of sale, supported mainly by tourism enterprises, have undertaken the distribution of the tourism product.

It is often crucial for the firms involved in the tourist industry to follow the pace of technological developments by gaining the necessary background knowledge and skills in order to support their investment decisions for the creation of their own e-Marketplace. This has also to be coupled with a continuous broadening of their perspectives based on a long-range strategic plan.

In the words of Michel Godet (2004): 'The real difficulty does not lie in the choice of proper solutions, but in making sure that everybody involved asks themselves the right questions. A problem properly framed and put to the people concerned is half solved already.'

So, in the context of the present competitive environment, tourism enterprises need to realign their current strategic approach, which is mainly driven by a short-term profit perspective and orient their strategies towards foresight and innovation. Recalling the words of Gaston Berger (1964), it is certain that 'if we look at the future, the needs of the present are set out before us.'

Tourism enterprises can be placed today in two broad categories on the basis of which activity contributes most to turnover. The first of these categories consists of companies which are involved in the organization of tours and packages, functioning as tour operators (outward- and inward-bound tourism). The second category consists of travel agencies, which, because of their small size, attempt to handle all aspects of the tourism product, functioning as 'tourism enterprises for general tourism'.

Members of the first category put together the tourism product from the primary materials drawn from the suppliers and then channel it towards the consumer. Members of the second category obtain their goods wholesale from the first group, which allows the members of the second group greater flexibility in methods of payment.

Local peculiarities are always of vital importance and serious note should be taken of them, since they shape the market trends accordingly. In Greece, for example, because of the small size of the market, economic resources and internal organization of businesses, members of the first category are very few, while the majority of businesses belong to the second group (*Greek Tourism 2010: Strategy and Goals*, SETE, 2003).

A closer look reveals other subgroups of enterprises intensively active in vertical markets, such as for example, businesses selling tickets of various kinds (IATA and non-IATA agencies), businesses specializing in conference organization, businesses specializing mainly in business travel or extreme sports, alternative tourism, and so forth. On the other hand, all tourism agencies depend directly on their supplier.

Tourism enterprises, on the supply side, such as hotel enterprises, small businesses engaged in renting out rooms, car-hire firms, airlines, ship companies engaged in coastal travel, taxi companies, tourist coach companies and businesses, e.g. restaurants dealing with catering, provide their services to the retailers. Thus retail tourism enterprises shape policy, regarding the distribution of the final tourism product, depending on how the commercial relationship between supplier and retailer is established.

The struggle for survival has led many enterprises to join the chain of electronic commerce, with the creation of their own electronic points of sale, that is, e-Marketplaces. This has been the response, not to their perfect understanding of the role of new technology and its advantages, but rather to the following factors:

- The introduction of new technology will reduce their overheads, since enterprises are aware that communication and printing costs for brochures, advertising leaflets, correspondence, etc. compose a rather important part of their total costs. It is remarkable that awareness is increasing among businesspeople that new technology will not reduce the number of jobs but, on the contrary, redistribute and enhance the productivity and potential of their workforce, for the general benefit of the company.
- Technology is enabling them to have a direct access to the customer, a vital issue for the viability of their business. A broad range of services are now available, such as: on-line info on the latest offers; better planning of new products, on-line changes in prices; and immediate response to client's needs with reliable information. Announcements regarding the few remaining seats on a trip can be made at the last minute or tickets can be auctioned, as it were. All this was, even in the recent past, just wishful thinking.
- The rapid reduction in costs of the hardware and software available to the home user, together with more user-friendly operating systems, has led to the misapprehension that: 'We, too, can do it now and certainly better than our competitors.'

All in all, businesspeople have realized that large amounts of advertising capital is required for disseminating information and that maintaining e-Marketplaces, with their continuously updated contents, is not an easy task. Finally, it has also become clear that hardware and software infrastructure is important and that rather specialized staff is required not only to maintain their operation, but also to combine sectoral knowledge on tourism with its own specialized knowledge, thus giving rise to the realization of training problems.

So far, attempts in the direction of e-Marketplaces seem to have failed, due to:

- inadequate internal business organization and strategy;
- the tendency of the tourism product suppliers to operate in the market without any foresight and planning, but rather on the basis on their personal assessment of the trends.

These failures have caused suppliers operating in the market to act as direct sellers of their services, while consumers, that is, travellers, have lost their confidence in the services offered by such e-Marketplaces. In this respect, both prices and the quality of the product are changing, based on the attitudes of the suppliers towards the market conditions. Therefore, the quality of the product is worsening to the detriment of the faith of the consumer in the market.

Such problems are growing, however, because e-Marketplaces suffer from a poor ICT infrastructure. The fast rate of new technology developments results in numerous platforms being used for the creation of e-Marketplaces, which are frequently exhibiting incompatibility problems, let alone those problems associated with the continuous upgrading and genuine structural changes required by technology.

Major investments have frequently come to grief because the programming tools employed to create the e-Marketplace in question were no longer supported by the information technology market. The companies that developed them had either closed or been bought out by competitors, who naturally wished to promote their own products to the group of clients they had just acquired. This, of course, is rather disastrous for a sector whose main problem is informing its clients immediately about offers and new prices that literally come about at the last moment.

On the other hand, moving the content of sites, already in digital form, from old to new platforms is a matter of very great importance. Such a move creates problems in the functioning of applications created at great effort and administrative expense. It also creates problems in the very structure of the information itself, so that such information requires reconstituting.

Things may become more complicated when the suppliers linked to e-Marketplace have to upgrade their own system infrastructure and train their personnel in ICTs. As a result the process of updating the content of the e-Marketplace, which is particularly costly, becomes a barrier for the successful operation of the e-Marketplace.

Another important matter is the question of the security of transactions. Banks and other fiscal organizations are the players who somehow set the rules of the game. Here are found some of the problems involved that require further attention, such as the use of electronic/digital signatures. In this context, the time required to respond and confirm financial transactions is vital for dispatching the customer's request.

The issue of competent management and technical staff is also important for e-Marketplaces. The management field is flooded with tools and approaches developed mainly in Japan and America. The utter collapse, however, of whole sectors of American industry in the 1960s and 1980s has cast some doubt on the classic American approach, which led 'managers to rediscover the virtues of positioning themselves against the best (benchmarking), the value of a complete overhaul of processes and structures (re-engineering), as well as the importance of sticking to the basics (downsizing) and, lastly, the power of innovation when it comes from the company's macro-competences' (Michel Godet, 2004).

It is the very complexity of the structure of the product in tourism and the need for certain quality that has imposed what was required all along, that is, the laying of the foundations of e-Travel Business in the proper manner.

5.3 Factors affecting the e-travel business chain

It would seem that the greater part of the product in tourism will continue for the foreseeable future to be sold via tourism enterprises in the same fashion as it is at present, that is manually. E-travel business is very far from being just a matter of forwarding and dispatching a client's request by e-mail.

The main reason why the transformation of the tourist industry as regards e-Travel has been so delayed is the limited access, in practical terms, to both the means of promotion and to focussed marketing, through the appropriate technology for reaching the consumer. There is also the important point that the decision-making process in tourism businesses is often rather confused, or as Jacques Lesourne (1989) puts it: 'Big decisions are rarely made. They become increasingly unlikely as long as small, everyday decisions pile up.'

The use of technologies such as digital TV, mobile phones and wireless networks, will constitute good means of disseminating messages regarding products and services, such as trips available at the last minute, ticket sales, tourist packages and special tourist events.

On the other hand, the basic players in the e-Travel Business will remain the airline companies. Thanks to the continuous improvement of their GDSs and the merging, at some point in the future, of their central reservation systems, such as Galileo, Sabre, Worldspan and Amadeus, these companies will continue to give direction to an important part of the industry. This is because the survival of tourism businesses, and particularly of IATA tourism agencies that also re-sell air tickets, as well as of their associates who depend on them, such as non-IATA agencies that buy tickets from the IATA agencies, depends on the policy pursued by the airline companies regarding the provision of commissions to these businesses.

In recent years a number of factors have led to a lengthy international recession in the travel sector – factors such as terrorism, wars, natural disasters, economic policies and changes in the geopolitical scene. These, in turn, have provoked a range of sharp reactions. Airlines have decreased their commissions to businesses; the tourism product itself has been restructured, with the economic survival of the suppliers chiefly in mind. The result has been that the consumer is now thoroughly confused and ends up buying services other than those offered to him (*A Year after 11-S: Climbing towards Recovery?*, World Tourism Organization, 2002).

The figures speak for themselves. In 2003, there was a worldwide decrease of 2 per cent in outgoing tourism. The reasons for this were the war in Iraq, the threat of terrorist action, the world recession and even the SARS epidemic. Locally, this decrease was dramatic, with Asia suffering a decline of 11 per cent and America 1 per cent, though Europe had a marginal increase of 0.5 per cent, the travellers in question here being Europeans themselves. Spain and France were the most popular destinations, with 3 per cent and 12 per cent of the market share respectively (*World Travel Trends 2003-4, WTM Travel Report*, IPK International, 2003).

For these tourism enterprises which consider the e-Travel Business as the way out of the vicious circle of the present times, there still remains open the important question of sustainable growth (*Structure, Performance and Competitiveness of European Tourism and its Enterprises*, EEC, 2003).

In the words of Hamel and Prahalad, 1994, the conclusion is as follows:

Some management teams will simply show more foresight than others. They will manage to imagine products and services and even entire sectors of economic activity that still do not exist and they will speed up their arrival. The sure thing is that they will not waste time wondering how to position their enterprise within the existing competitive environment, because they have already created new environments. The other enterprises, the so-called 'laggards' will worry more about preserving the past than conquering the future.

The GDSs, the large-scale, worldwide networks of the airline companies, rest upon already outmoded technologies. These networks were developed approximately three decades ago, and are hosted either on their own technological systems, or on systems provided by the ICT industry of today.

The e-Marketplace is a basic element in the architecture of the e-Travel Business. When tourism enterprises began to develop such systems, it became clear that these should be linked with the GDS platforms. They therefore developed a range of special applications that function as booking engines.

A process thus started that has resulted in the creation of the many links in the chain of e-Travel (see Figure 5.1). Among these links are Databases of Site Content (e.g. information on hotels, complete with description and photographs), or Customer Content (website registration), or Engines designed for Searching for Prices and Costing directly from the Web itself.

Other basic elements in this context are Back Office Applications for handling the internal organization of enterprises, thereby attempting to bridge the gap between all the administrative details of the enterprise and the e-Marketplace on the Web.

A crucial link in this chain is the set of processes for payment on-line, by credit card, determined by the banks. The consumer who intends to travel visits the e-Marketplace and purchases the chosen service by using a credit card, a procedure that is secure in terms of safety and validity.

5.4 E-Marketplaces versus tourism product suppliers: How to solve the riddle?

The tourist product is, by its nature, complex and multidimensional, although it would also seem that this complexity cannot be regularized, since tourism is a sector sensitive to a range of changes arising from conditions both at the local and international level.

A valid example illustrating this sensitivity concerns the preparation of a package by a tour operator. In this case it is of vital importance that the bookings at the hotel should be perfectly coordinated with the times of the charter flight to and from the destination.

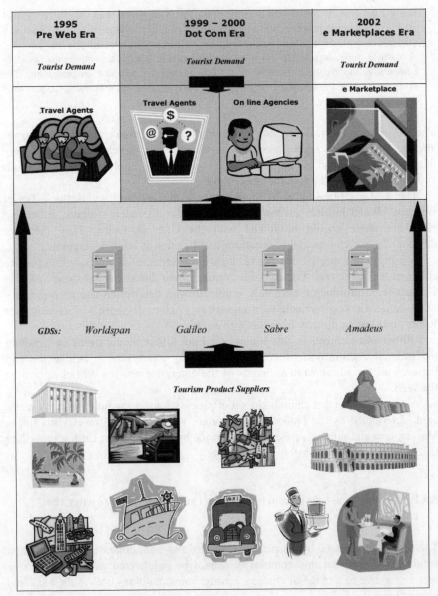

Figure 5.1 The e-Travel Business Chain

Tour operators frequently pre-purchase and pre-pay the suppliers for their services very much in advance of the tour. Small hotels are very eager to offer their rooms at very low prices, as long as payment and pre-payment are guaranteed. Thus the only person at risk is the tour operator. When unforeseen circumstances, such as fluctuations in exchange rates, bad weather conditions or changes in the economy, drastically influence demand, operators are compelled to offer large discounts, which result in serious loss of revenue. It is then that last minute e-Marketplaces, such as lastminute.com, can undertake to rescue whatever possible.

In the case of an e-Marketplace that belongs to a tourism enterprise, all the relevant information needs to be prepared for entry into the system first by a technical support mechanism and then immediately placed in the system before being made public. The point, however, that the whole process depends on the supplier becomes of more importance in simple, daily matters. But there are no rules governing the frequent changes in the suppliers' prices. These unjustifiable changes are excused by the operators on the grounds that the product is sensitive.

Matters become even more complicated, when suppliers operate directly as tour operators, that is, they bypass the chain and offer their services directly to the consumer, thus destroying the reliability of the whole procedure. The consumer, who has saved for the whole year in order to be able to afford this holiday, is left without knowing who will protect him or who will guarantee the quality of the service that is to be purchased.

Retail businesses form the vanguard of the industry vis-à-vis the consumer. The foundations of the chain of commerce are shaken when such businesses have purchased a particular service, via a special agreement, from the e-Marketplace, and then discover that the process cannot be relied on. The cost to the business in question is considerable. They have invested in printed material and in its dissemination and they have invested in mobilizing resources to reach the customer. Likewise, the cost that they have shouldered in order to publicize their e-Marketplace to the consumer is considerable.

Such conflict occurs, then, mainly at the level of wholesale transactions (B2B trade). At this point, an opportunity for a better organization of the system is required as well as the creation of a regulatory framework of operation.

The structural element in the chain is the e-Marketplace. This may be regarded as a central node in a Small Closed Network (see Figure 5.2). On the one side stand the suppliers of on-line services, while on the other stand businesses engaged in reselling.

It is thus difficult to plan an e-Marketplace, although it can be done. The suppliers in such a schema also supply other small closed networks at the same time. The difference lies in the peculiarity of the final product. So many years of successful handling by tourism businesses have ensured that this product has successfully given such businesses the reputation for specialization. This results in the discovery by the client base of the quality inherent in the product.

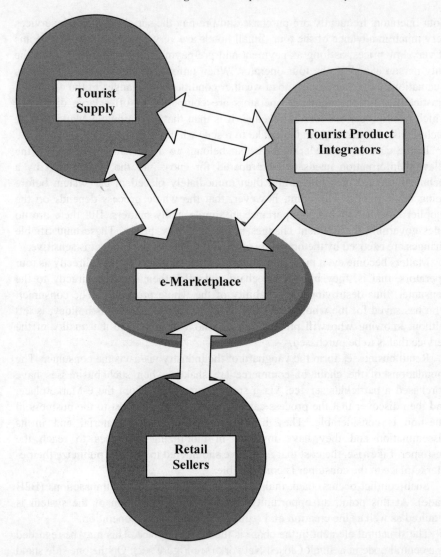

Figure 5.2　　The closed network

On the other hand, a small network offers a true educational process to its members who come to recognize the obvious benefits of participation: the minimization of communication operating costs; minimization of printing costs; accurate information; speedy delivery; direct communication with collaborators and suppliers; and an impressive on-line product presence in the eyes of the customer.

　　Tourism will continue to be viable, as long as both the agents of distribution and supply of the product and tourism enterprises themselves choose to implement e-Marketplaces for B2B travel. This is to be done by means of Closed Private

Networks in order that suppliers, institutions and collaborators can work together, so that they will then be able to conquer the field of B2C e-Travel via focussed marketing.

5.5 Conclusions and prospects

In the present chapter, the working assumption has been that Business-to-Business (B2B) e-Travel should be implemented via Small Closed Networks. The elements of the system are the tourist product Suppliers; the tourist product Integrators; the retail Sellers; and, finally, the e-Marketplace. The e-Marketplace element operates as the 'controller' of the system's elements interrelations.

On-line sales in B2B e-Travel have spread to many private e-Market places run by travel agencies. However, the attempt by a few e-Marketplaces to dominate the product distribution chain has failed dramatically, because of the suppliers' insistence on maintaining an irregularly functioning administrative control over the product. For e-Marketplaces to survive and thrive means complete control over the primary services, offered by the tourism product supplier, at its source.

The major problem in the system's operation is that Tourism Product Suppliers act as Direct Operators. The complexity of the tourist product requires as the first step that the e-Marketplace administrators should undertake the task to lay the foundations of B2B travel. Thus the 'controller' element should absolutely define and sustain the system's element interrelations.

On the other hand e-Marketplaces suffer from poor IT infrastructure and a lack of qualified executives and managers. The use of technology, such as digital television, mobile telephones and wireless networks, will prove to be a good method of disseminating publicity regarding trips available at the last minute, tickets, tour packages and special tour events.

E-Marketplaces need to identify the most appropriate opportunities of these technology wonders, since no single platform will dominate. They must integrate on-line and off-line channels to present consistent and up-to-date details of the tourism product. They should also give serious consideration to the fact that the airline industry will still be the major mentor in the e-business game, via the development and merging of its central reservation systems.

To speed up the e-transformation of the tourism industry, promotion of e-Marketplaces with focussed marketing must take place, guided by the system's key role players.

References

Association of Greek Tourist Enterprises (SETE) (2003), *Greek Tourism 2010 – Strategy and Goals*, (2nd edn).
Association of Greek Tourist Enterprises (SETE) (2003), *Tourism and Employment*.

Berger, G. (1967), *Etapes de la prospective*, PUF.

Boyer, L. and N. Equilbey (1990), *Histoire du management*, Editions d'organisation.

European Economic Community (2003), *Structure, Performance and Competitiveness of European Tourism and its Enterprises*, Directorate General for Enterprise.

Godet, M. (1997), 'Manuel de prospective strategique', vol. 2, *L'art et la methode*, Dunod, Paris.

Hamel, G. and C.K. Prahalad (1994), *Competing for the future*, Harvard Business School Press, Boston, Massachusetts.

IPK International (2003), *World Travel Trends 2003-2004*, Forecast Forum: WTM Global Travel Report.

IPK International (2004), *World Travel Monitor Findings*, ITB, 15 March, Berlin.

Lesourne, J. (1989), 'Plaidoyer pour une reserche en prospective', *Futuribles*, no. 137, Nov.

Rayan, Ch. (1990), *Recreational Tourism*, Routledge, London.

World Tourism Organization (WTO) (2002-3), *A Year after 11-S: Climbing towards Recovery*, News from the World Tourism Organization.

Chapter 6

ICTs and Local Tourist Development in Peripheral Regions

Anastasia Stratigea and Maria Giaoutzi

6.1 Introduction

Information technology and its applications provide a broad range of challenges touching upon many fields of human activity. Many researchers claim that their integration into firms' activities as well as their evolving synergies will radically *re-engine* many processes in the course of the next decades (Buhalis, 1997).

ICTs are *re-engineering all tourist-related processes* by supporting new business models, restructuring tourist distribution networks, providing new innovative marketing strategies, etc. The impacts of the above changes are profound during the creation process of the economic value of the sector. On the basis of these innovative processes, the tourism market place is gradually being replaced by the tourism market space, where infrastructure, context and content constitute elements of paramount importance (Hudson, 1997; Kunz, 1998).

The main value-adding characteristic of ICTs in tourism lies in their potential to provide *access* to both the demand and supply side of the tourist sector. Actors on the demand side can gain access to tourist information mainly through the World Wide Web. Suppliers of tourist products on the other hand – businesses and places – can enhance their potential to manage and disseminate information by developing new aspects in distribution and sales channels for all the tourist products.

In remote peripheral areas, such access increases their competitiveness in the tourist market, by strengthening their position in keeping control of the type of the tourist product they would like to promote. As a result, an expanding *geography of tourist resorts* is developing, where new tourist resorts are constantly emerging and compete successfully on the basis of their specific tourist product. Under such a regime, every place in the world may potentially become a tourist destination (see Talero and Gaudette, 1995).

Hence, remote and peripheral areas, willing to compete on the grounds of the newly emerging tourist market space, are forced to focus on the adoption and use of innovative strategies in order to cope with the increasing market competition.

This will enable them to acquire the critical mass of know-how and technology needed for the support and strengthening of their position in the tourist market. Training in ICTs and increased awareness of the capabilities of new technologies, Web presence and cooperation in a network context for sharing resources and opportunities will all enable local firms to re-engineer tourist business processes and compete with the large tourist firms at present exploiting these areas.

New business organizational schemes such as *virtual organizations* are innovative strategies, characterized by their product-market strategy, network structure, information systems and business communication patterns. These may constitute a strategic 'tool' especially for small firms in remote and peripheral areas, in order to cope with the increasing competition and ensure efficient use of pool resources.

The present chapter focuses on the potential role of virtual organization structures on local tourist development, with emphasis on comprehensive tools supporting local tourist development in remote and peripheral areas. Section 6.2 presents contemporary trends in the tourist sector by putting emphasis on all aspects of the tourist sector affected by ICTs. Section 6.3 elaborates on the contribution of virtual organization structures in improving the competitive position of tourist SMEs located in peripheral areas at a global scale. Section 6.4 further explores comprehensive tools based on ICTs, which support local tourist development in peripheral areas. Finally, Section 6.5 draws conclusions and points out further research issues.

6.2 The tourist sector and ICTs

The scope of this chapter is to present contemporary trends in the tourist sector in the light of ICTs. Therefore it focusses attention on the impacts of ICTs on the tourist market, by putting emphasis on demand and supply aspects and the re-engineering processes in the tourist sector.

Tourism is a *structural element* of modern societies. Its contribution to local economic development is presented in many empirical studies, while its future development is expected to be tremendous. As Naisbitt (1995) claims, the global economy in the next century will be driven by three sectors: information technology, telecommunications, and tourism.

Many researchers have stressed the role of ICTs in the tourist sector in the global marketplace (Mutch, 1995; Buhalis, 1998; Kunz, 1998; Connell and Reynolds, 1999; Buhalis and Schertler, 1999). They claim that specific characteristics of the sector are:

- The *international focus of the tourist business* based on the nature of the tourist activity itself and the small size of the domestic tourist markets (Chetty and Wilson, 2003). This entails very specific characteristics of the tourist sector, which has to cope with the large distances usually involved between origin and destination, the time-sensitive transactions, and multiple trading partners.

- The *intangible nature of the tourist product* that is usually sold before use and away from the place of consumption (Buhalis, 1998). Selling the product depends, among other things, upon its attractive presentation. ICTs in this respect are offering a great potential with regards to the worldwide, timely, up-to-date information provision, relevant to the consumer's needs. They actually provide the information backbone, which will establish on-line interaction between origin (potential travellers) and destination (tourist resort or tourist business). Swarbrooke (1996) identifies various applications of ICTs, such as the Global Distribution Systems (GDSs), Computer Reservation Systems (GRSs), Destination Management Systems (DMSs), Multimedia Systems, Interactive Televisions, smart cards, electronic mail, computerized management systems, etc. These are considered as major technological developments for tourist information handling.
- The *heavy consumption of a diverse range of information* (Cho, 1998; Sheldom, 1997; Archdale, 1993). ICTs provide a particular seedbed for applications which will ensure efficient management of tourist information, high-quality delivery, speed, accuracy and responsiveness. As evidence shows, the tourist sector is at the forefront of the ICT users (Connell and Reynolds, 1999; Mutch, 1995).

Also stressed by many researchers is the power of ICTs in *reshaping the nature of competition* in the tourist sector by directly linking consumers and suppliers, thus adding value to organizations' products (Buhalis, 1998; Tsolakidis, 1998; Buhalis and Schertler, 1999; etc.). The outcome of the above process has been reflected on both the demand and the supply side of the tourist sector.

On the *demand side*, ICTs support the concept of the '*mature consumer*', that is a new, sophisticated, knowledgeable and demanding consumer, very familiar with ICTs, who seeks flexible, specialized, accessible, interactive products and communication (Buhalis, 1998). His direct interaction with the providers of the tourist product increases his skill, knowledge and transparency of the options available, thus facilitating his choice.

On the *supply side*, ICTs enhance the innovative potential of firms in the tourist market by contributing the effective management and distribution of knowledge and information, direct access to the consumer, influence of the consumer's behaviour through proper destination marketing systems, re-engineering of business processes, promotion of SMEs in the tourist sector, etc. (Buhalis and Schertler, 1999). ICTs can also offer to the tourist firm new management and business opportunities, which may affect them in *at least four different ways*, (Buhalis, 1998), by:

- enhancing their competitive advantage in the global tourist market;
- improving productivity and performance;

- introducing new ways of management and organization;
- developing new business opportunities.

The use of ICTs has altered the processes in the tourist sector, as follows:

- The *globalization of the tourist markets* is greatly supported and fostered by advances in the information technology, thus enabling an increase in the market range from local to global (Wayne, 1998; Tacgmintzis, 1999).
- The *change of tourist players and processes* was marked by the entrance of new companies and intermediaries as well as by the access to new innovative direct management and marketing strategies, which influence destination management and distribution (Tacgmintzis, 1999; Wayne, 1998; Kunz, 1998).
- The *customization of the tourist products* as tools for detecting and enlarging tourist segments, as well as for promoting specific tourist products to market niches through electronic distribution channels (Buhalis and Schertler, 1999).

ICTs are to a great extent facilitating and fostering processes towards the *globalization* of economies, markets, systems and cultures. Applied to the tourist sector, the trend exhibits two different paths. On the one hand, large tourist *multinational and transnational corporations* in travel and tourism are continuing to grow, on the basis of their economies of scale and scope. This strongly contributes to their penetration in the world tourist markets and the catering for a large volume of tourist destinations and movements (mainly in mass tourism). This also determines the future development of the tourist resorts around the world. Examples of such corporations are hotels, airlines, etc. (Lafferty and van Fossen, 2001).

Local operators, on the other hand, on the basis of their cultural and natural resources, are striving to survive and further develop their own specific market niches in order to promote their tourist products and services, oriented towards an *alternative* tourist experience. Their comparative advantage lies in the growing trend for *alternative tourist paths* (Stamboulis and Skayannis, 2003).

Apparently an *inevitable friction between global and local forces* is developing, where the tourist sector clearly shows the signs of such a friction, mirroring in its structure the global local polarization processes (Wayne, 1998).

In this context, the adoption of ICTs may affect the *position of the tourist players* in the marketplace, thus reshaping the geography of power relationships, where the position of local tourist SMEs can be reinforced.

As information becomes the source of power in the information era, bargaining relationships among partners are reshaped and the power is gradually relocated from institutional buyers and wholesalers to suppliers (Buhalis, 1998), because of their potential to establish *direct access* to customers. This may create substantial benefits for both customers and suppliers – the win-win orientation as expressed by Fischer (1998). Thus effective and interactive communication between local tourist SMEs and their target markets, becomes a crucial point for their future development and competitiveness.

In such a context, tourist SMEs, on the basis of their small size, coupled with innovation and effective networking, are able to 'develop and deliver the right product, to the right consumer, at the right price and place, without over-dependence on intermediaries' (Buhalis, 1998).

The market share is not very clear-cut among the various tourist providers (Kunz, 1998; Buhalis, 1998). At the same time, tourist demand is increasing both in size and complexity. More and more people are seeking a completely diversified tourist experience. Moreover, sophistication of tourist products is also rapidly increasing, thus leading to specific market segments sharing certain characteristics, attitudes and behaviour. There appears to be a growing trend towards the *customization* of the tourist product based on the use of electronic technology, which will establish the potential of a 'one-to-one' marketing process.

As evidence shows, ICTs are the proper tools for the identification of *tourist segments*. They also increase the effectiveness of *customer service* by providing new services or improving existing ones. Various applications appear to be of great help in the effort made by the tourist firms to detect and enlarge such market segments, thus providing the opportunity of 'thinking global while acting local' (Wayne, 1998). Such applications may constitute effective 'tools' towards *destination management and distribution*, on behalf of both tourist businesses in their efforts to promote a specific tourist product, as well as attracting customers in their search for alternative tourist choices.

The *consumer behaviour* towards tourist products has also been greatly affected by the use of ICTs, where options are greatly enhanced by the ability to search for transparency of price and variety of offer of tourist products. Knowledge, maturity and transparency characterize consumer choice, through direct access to the tourist options available via ICTs, which are the keystone of this effort. Moreover, ICTs and especially the World Wide Web provide the opportunity for inexpensive delivery of multimedia information of both tourist products and destinations. A wide variety of tailor-made tourist products are at the consumer's disposal, which has further enhanced his range of options, from the totally 'comfort-based' demand to the 'adventure' or 'education' demand of the tourist product, thus supporting the polarization of tourist taste on a global scale (Wayne, 1998).

6.3 Virtual organization in local tourist development

The scope of this chapter is to elaborate on the prospects introduced by virtual organization structures for local tourist development. The focus therefore is on competitiveness aspects of the tourist SMEs in peripheral areas and ways that virtual organization structures may support SMEs in coping with the new challenges and opportunities in the globalization era.

Peripheral areas in the globalization era have to cope with both threats and prospects affecting tourist development potential. Threats are related to the increasing pressure exerted upon them by the increasing competition in the sector. Prospects, on the other hand, are associated with the potential offered by ICTs to be more competitive and to search for new opportunities. Andrews (1971) notices that there is a need for strategic action of tourist firms in peripheral areas, which usually is the outcome of the firms' existing competence and the availability of new opportunities.

The fact that the tourist product may be the basis for a successful tourist development in a region is not sufficient in itself. The potential tourist of today is more demanding and sophisticated and requests high quality and a range of products, as well as direct interaction with the supplier in order to satisfy specific needs and wishes. Future perspectives of the tourist sector in peripheral regions lie in the timely identification of consumer needs and the ability to respond with comprehensive, personalized and up-to-date information to the potential demand (WTO, 1998).

The issues of *short term response* to the tourist market demand and of *flexible adaptation* to the customer needs both stress the need to incorporate technology, enhance direct interaction with the marketplace and adopt flexible *business organization schemes* that enhance cooperation within the tourism industry and promote globalization (Buhalis, 1998).

The *virtual corporation* concept is evolving as a specific example of technology-driven flexible networked organization. It actually represents a *new business strategy*, which is heavily based on ICTs, aiming at the realization of short-term flexible cooperation schemes, based on information and knowledge sharing, that are structured both horizontally and vertically (Jagers et al., 1998). Its strength lies in the *integration* of different parts/enterprises (the best-in-class-network) in schemes with a high degree of horizontal structuring, low formalism and strong team orientation.

Virtual organization provides both great potential for the strengthening of the position of tourist SMEs in a globally changing competitive environment, and a prominent structure for their future pervasiveness in the world tourist market. It actually represents a new quality of economic value creation in the tourist sector, putting priority on 'thinking in core competencies' instead of a 'cost-driven strategy' (Fischer, 1998). It affects tourist businesses since it:

- increases *efficiency* of virtual schemes by sourcing intellectual *resources* globally (Skyrme, 1998);
- increases *global presence* without local presence;
- increases *flexibility* through dynamic structures and contractual agreements, as well as exchangeability of networking partners according to the problem at hand;
- assures *higher quality* and *range* in the tourist products;

- increases the potential to tackle *projects* or *problems*, which might otherwise have been beyond their capabilities (Skyrme, 1998);
- creates substantial benefits for all parts of virtual partnerships – the win-win orientation of virtual schemes (Fischer, 1998).

Virtual organizations in the tourist sector may offer considerable support to *management* and *business strategy* and can be established both at the *intra-organizational* and *inter-organizational level*.

The *intra-organizational perspective* relates to virtual structures between functions/branches of a certain tourist firm (vertical structure). ICTs in such a context enhance the level of integration of a number of intra-organizational processes, e.g. accounting, marketing, customer service, etc. The outcome of this integration relates to an increase in efficiency and productivity of the firm, as well as to effective operational management.

The whole interaction among different functions within a tourist firm is taking place by use of *intranet technology*, i.e. an internal network deploying the same technology and presentation tools as the Internet, giving access only to authorized personnel.

The *inter-organizational perspective*, on the other hand, relates to the *networking* among either *tourist firms* or *tourist destinations* at the local, regional or even global level. Its power mainly focuses on the support of the tourist firm in achieving strategic goals.

Networking among firms reflects the multi-sectoral nature of the tourist sector, where tourist firms are cooperating on the basis of pooling resources for efficient planning, production, distribution and selling of the tourist product.

Networking among tourist destinations is the other important perspective of inter-organizational virtual schemes. Such cooperation may take place at a regional or inter-regional spatial level. This type of cooperation is greatly supported by the current polarization trend in tourist behaviour, enhanced by ICTs. It may result in *multilevel regional cooperation* in the tourist sector, which leads to mutual satisfaction of the various sides, but also more attractive tourist product attributes, a richer and more comprehensive tourist offer, larger market segments, etc., thus optimizing value-added processes.

The concept of inter-regional cooperation may apply to various spatial levels, ranging from the small region, in its effort to integrate various tourist services into a more rich and diversified tourist product, to the inter-regional or even international level.

Extranets in this context are providing a secure framework for the network organization perspective, ensuring access to authorized organizations (Buhalis, 1998).

6.4 Local tourist development for peripheral areas

The present chapter deals with issues which relate to the potential offered by *virtual organizations structures* to *local tourist development* in *remote* and *peripheral areas*. More precisely, it focuses attention on the specific characteristics of peripheral areas as tourist destinations and the role of ICTs and virtual organizations (VOs), in promoting tourist assets in a global context. The issue of virtual organizations is approached on the basis of: Destination Management Systems (DMSs) as virtual organization among stakeholders at the local level – peripheral areas (first degree of virtualness); virtual organization among DMSs at the inter-regional level (second degree of virtualness); and virtual organization among firms located in peripheral areas.

Peripheral areas are generally areas lagging behind as to their development, due to a range of factors such as political processes; available resources; development patterns; production factors of limited size and diversity; network infrastructure and its performance, etc. As a result, they are characterized by poor investments, low income, high unemployment rates, low population density, etc.

These areas, on the other hand, are exhibiting all the characteristics which mark them as interesting sites for tourist destinations. Their comparative advantages lie on their *natural and cultural resources*, which when properly handled may constitute a tourist product of high quality. The effective promotion of this product, in the context of sustainable tourist development, may have considerable implications for their future development perspective.

In order to cope with the increasing competition in the tourist sector in the broadening global context, peripheral areas need to develop a *focussed, specialized product*, which will be based on their physical assets and is relevant to the social, cultural and environmental attributes of their local communities. At the same time, in order to *properly communicate* this product worldwide, they need to gain access to, and make efficient use of, technological resources.

As already noticed above there is a clear shift in consumer's attitude towards tourist activity, exhibiting clearly the signs of a *mature consumer* seeking an *alternative tourist experience* based on the use of ICTs.

Remote and peripheral areas are in an advantageous position in this respect, provided that they 'get the message' and take advantage of their human and natural resources towards the satisfaction of this demand. Apart from the product supplied, a crucial aspect is the *on-line presence*, being the 'meeting point' between the mature consumer and tourist development of peripheral areas.

In this context, peripheral areas need to adopt a *new strategic approach*, which should be both *product-focused* and *promotion-focused*.

Focus on product implies the creation of a proper destination image, which will differentiate the tourist product of a certain peripheral region from that of its competitors. This should be the outcome of a participatory process among all stakeholders in the area of concern, aiming at the development of peripheral areas as tourist destinations. *Focus on promotion* on the other hand implies access to, and use of, those distribution channels, based on ICTs and their applications, which will support the distribution of this image to the right audience worldwide: namely, market niches.

A new strategic approach in the sector calls for the use of *comprehensive tools* based on ICTs and their applications which will ensure *on-line presence* and support the *direct interaction* with customers and buyers of their tourist product. These may apply at both the *macro-level* (namely, *peripheral areas*) and the *micro-level* (i.e. *the tourist firms*). Figure 6.1 presents the role of such specific tools based on ICTs which are used for tourist development in peripheral areas at both levels.

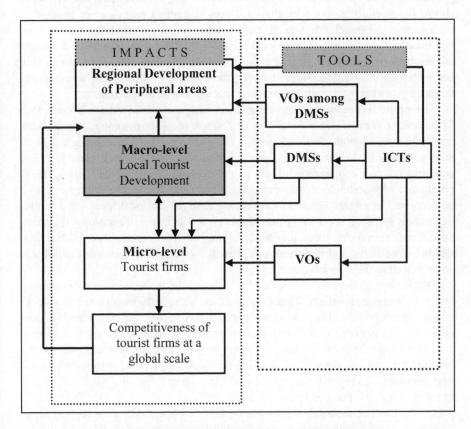

Figure 6.1 ICTs for local tourist development in peripheral areas

The new strategy and its basic components – tourist product and promotion – bring to the fore the concepts of *tourist destination* and the tools for *tourist destination management* as core issues in the efforts of peripheral areas for regional development in general, and local tourist development in particular.

Several critical aspects concerning local tourist development in peripheral areas with regard to the tools offered by ICTs and their ability to *strengthen advantages* and *alleviate disadvantages* in such areas are presented here:

The first aspect relates to the *planning of the tourist product* that peripheral areas wish to promote. This is a very important task for these areas, since the 'product', although demand-driven, should take into account socio-economic, environmental and cultural aspects of the area of concern along the lines of sustainable development. Each tourist destination is conceived as a total of tangible elements, such as products, services, facilities, local attractions, and intangible elements, such as culture, local customs and traditions (Buhalis, 2003). The tourist product as the image produced by the synthesis of tangible and intangible elements should be carefully planned by the local region through participatory processes, incorporating all stakeholders in the area: namely, firms and the local society. The key issue in such a context is that local societies should *decide* on and *keep control* of, the type of tourist development they wish to promote. The outcome of this process is the *destination image*: namely, the collection of products, services and facilities provided by all stakeholders in a specific tourist destination.

The proper promotion of the destination image is based on the *Destination Management Systems* (DMSs), which are strategic tools for planning, management and marketing of both *destinations* and *local enterprises*. Their strength lies in their potential for *destination integration* (Buhalis, 2003), which calls for high coordination and cooperation of all parts involved: namely, the region, local providers and stakeholders in a *network context*. DMSs in such a framework may be perceived as *virtual organizations* among firms at the local level – the tourist destination – using *intranet* (within DMSs) and *extranet* technology (between DMSs and firms) for local tourist development. The outcome of DMSs is a flexible, specialized and integrated tool, benefiting *regions* and *firms* in peripheral areas as well as their *customers*.

DMSs support *peripheral regions* by assuring the maximization of multiplier effects in a regional context. They increase their competitiveness in terms of local tourist development. They support participatory processes in planning and managing the system, by integrating all actors at the local level: namely, tourist and non-tourist players. This in turn implies entrepreneurial and social participation and coordination in defining the *tourist identity* of peripheral regions (the product), giving the chance of keeping control on the kind of tourist development that it pursued in those regions.

On the basis of their *destination integration function*, DMSs can support long-term prosperity, competitiveness, and the development of peripheral areas as host areas, by increasing economic output in all sectors and capitalizing on multi-integration of local partners at the regional context (Buhalis, 2003). This has positive impacts on the local economy in peripheral areas, thus reversing the economic climate in terms of attraction of investments, income, employment rates, population restraint, recruitment of highly skilled labour, etc., which are the weak elements of peripheral economies.

In the hands of public agencies, DMSs can also greatly support *policy interventions* aimed at integrated and sustainable socio-economic development of peripheral areas as tourist destinations. This stems from their power to provide a *platform* for SME support in terms of infrastructure and on-line presence, thus removing locational constraints and giving SMEs the chance to promote local assets on a global scale. This also stems from their power to integrate all sectors of economic activity into the final product. DMSs support worldwide communication and promotion of both the destination image and local SMEs and consist of the proper tools for market research, searching for *new opportunities* in the tourist market, which will enhance market niches of peripheral areas and contribute to the flourishing of both regions and firms. They can, for example, respond to a specific demand, e.g. sports tourism, by developing dedicated parts of their promotion policy to this type of tourist activity in a specific area.

DMSs can be led by public or private agencies or a combination of both. Practical experience shows that the majority of DMSs are led by public agencies, while some of them are led by public/private partnerships. In respect to peripheral areas, it is essential that leadership should be in the hands of regional or local public agencies or even national tourist organizations. Public-driven DMSs can play a crucial role in the planning process of such a system, by coordinating actions and local stakeholders towards sustainable local tourist development. These may also play a decisive role in running the system (content, infrastructure, networking of stakeholders, human resources, on-line presence), its smooth functioning and the continuous updating and enrichment of information provided, and the credibility of information, as well as in the investments involved on infrastructure and know-how.

In peripheral areas, DMSs support the competitiveness of *local firms*, by providing a variety of services to local entrepreneurs, by acting as the umbrella of technological infrastructure and human resources of high quality, and by making a range of functions and services available. *Firms* in this respect assure an on-line presence and access to resources and know-how thus reducing operating costs; take advantage of new opportunities established through DMSs market research; get access to specific functions/services provided by the system; reach the global market at an affordable cost; and reap the benefits emerging from a destination's reputation and image communicated by such a system.

Finally, DMSs can enhance the *customers' experience* concerning a tourist destination, giving them the opportunity to create, through the various options available, their specific tailor-made product according to their preferences, time schedule and wishes, thus developing individual tourist packages and making the necessary reservations *on-line*. This is quite important in the case of the mature consumer seeking alternative tourist experiences in quiet and peaceful local environments.

The *virtualization concept* may apply to *DMSs* as well. Such a framework implies a network cooperation of different tourist destinations in order to strengthen their market position and join forces to offer a rich and diversified tourist product. It may apply at a regional, interregional and national level according to its purpose. DMSs in peripheral regions may also form virtual corporations with DMSs of other regions, either in the context of *common tourist products and interests*, e.g. the Alpine regions (Fischer, 1998), or in the context of *diversified tourist products,* thus broadening the range and diversity of their offer, e.g. islands in the Aegean or the Mediterranean region. DMSs provide remote and peripheral areas with a unique opportunity to develop tailor-made services focussing on resources such as their landscape, lifestyles, cultural and environmental assets, which all combined to create a rich and attractive tourist product. Within such virtual structures, apart from gaining power in the tourist market, it is also possible to get cheap tourist destinations and exclusive and first-class advertisements, accessible to an enormous number of potential tourists (Tuntev, 1998).

The emerging virtual schemes may share, among other things, technology, know-how, information, efforts and resources in order to increase their market share worldwide. They strengthen interregional cooperation and complementarity not only in terms of tourist activity but also in general. They add value to the value chain of the tourist product and join efforts to explore in a more systematic and comprehensive way market niches for their products and services, thus increasing their share and competitive position in the global tourist market.

The role of DMSs in the process of tourist development in peripheral areas is crucial, as they constitute tools for the *promotion and marketing* of the destination image, and act as *platforms* incorporating and communicating any kind of tourist information related to destinations. The development of such systems and on-line communication of their content is very important to meet consumer's needs. Provision of such information through DMSs in conjunction with its content quality and diversity reinforces the *knowledge* and *attraction* of the potential tourist towards a certain tourist destination, increasing thus its competitiveness.

The importance of *DMSs* as *strategic tools for local tourist development in peripheral areas* are critical at both the macro- (destination) and micro- (firm) level, since:

- They entail *wide participation* and *high involvement* of *local firms*, increasing *co-operation* and *co-responsibility* among all stakeholders concerning the type of tourist product they wish to promote.
- They promote *integration* of all kinds of local products in the tourist product, thus integrating economic activities in peripheral areas. In this respect, they provide a *tool* for the regional development of peripheral areas.
- They promote a close *cooperation and ties* among firms and the region in terms of information interaction on opportunities, products, services, rates, tourist packages, etc. This results in a continuous updating and on-line communication of the system content, which in turn affects the competitiveness of the peripheral areas as tourist destinations.

- They ensure *accuracy, reliability, neutrality* and *quality of information* in order to protect consumers and local SMEs as well as the region's reputation and credibility, provided that they are under public control.
- They *sustain local SMEs* by providing access to resources, a quite important asset for such areas, which suffer from lack or limited size and diversity of resources.
- They support *interoperability among* all stakeholders/firms in order to enable them to operate in a network context.
- They assure unimpeded worldwide presence, by operating on a wide *variety of technological platforms* continuously conforming to the latest technological standards.
- They enable peripheral areas to promote their assets and compete on the grounds of quality and diversity of their tourist product, in the global tourist market, by *limiting their dependence on large tour operators* and controlling tourist flows.

At the *micro-level*, virtual organizations (VOs) (namely, firm cooperation among 'best-of-class' firms based on the use of ICTs in a network context) are becoming the 'heart' of firms' *corporate strategy*. ICTs as technological enablers are establishing an info-space, within which the whole tourist industry can operate and establish worldwide partnerships with suppliers, customers and intermediaries (Buhalis, 2003). Internet, extranet and intranet technology constitute the communication technology used, depending on the type of interaction.

In contrast to DMSs, which are networks developed among local enterprises in peripheral areas aiming at the construction and promotion of the tourist image involved, VOs among firms are cooperation structures oriented towards a certain goal. Success of the goal set is based on heavy or exclusive use of ICTs, which may be established at a local, regional or interregional level. VOs are offering the firms' flexibility and quick responsiveness in a continuously changing tourist market place, by incorporating the rapidly evolving consumer needs in their business strategy. They serve the purpose of sharing risks, responsibility, information and resources among firms towards a certain goal.

VOs in the tourist sector are extremely important for tourist SMEs located in remote and peripheral areas. These areas are characterized by production factors of limited size and diversity and limited economies of scale. Small tourist firms in such a context have no access to certain services at the local level. The virtuality concept enables them to acquire these services by outsourcing, that is, by forming VOs with, e.g. accounting systems, management and strategic information systems, points of sale of their product, etc. Tourist SMEs in remote and peripheral areas may remain competitive through proper networking.

From this discussion, it follows that *small tourist firms* of today located in remote peripheral areas should respond to the challenges provided by ICTs as organizational and promotion tools in a *multidimensional way* (Figure 6.2). In such a process, SMEs may constitute the *'heart' of local tourism development* in peripheral areas, by acting both:

- as *private enterprises* aiming at their own benefit from running the business by adopting VO structures either in an intra-organizational or in an inter-organizational context;
- as *parts of regional environments*, joining the image in a tourist destination context.

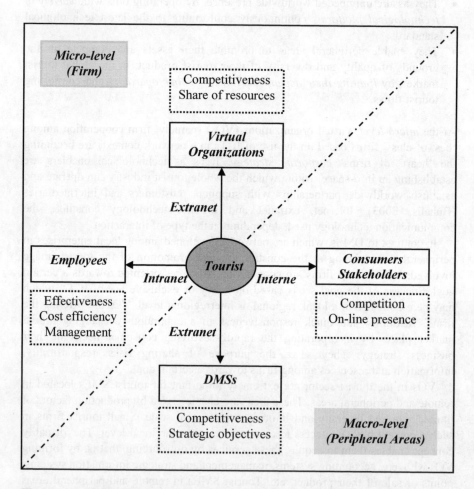

Figure 6.2   Multi-dimensional strategy of tourist firms based on ICTs
Source: Adapted from Buhalis, 2003.

Long-term planning and strategy, re-engineering of business processes, top management commitment, and continuous education and training throughout the business hierarchy (Buhalis, 1998) in the context of ICTs, all seem to be essential elements in the multidimensional strategy of tourist firms.

6.5 Conclusions

The development of ICTs and their introduction in the tourist sector have strongly influenced the nature and scale of competition on a global scale. Under the new circumstances, SMEs are facing new challenges with respect to their dynamic entry in a continuously changing tourist market. Peripheral areas, on the other hand, are competing on the ground of the alternative tourist paths they offer, taking advantage of the shift of consumers' behaviour towards new patterns of relaxation and entertainment.

In such a framework, SMEs in peripheral areas need to adopt a multidimensional strategy, the heart of which is cooperation at both a micro- and a macro-level. This cooperation is based on new organizational and innovative promotion structures.

Remote and peripheral areas, as well as tourist SMEs located in such areas, have a great chance to promote their tourist assets and products by use of comprehensive tools based on ICTs. In order to succeed in their efforts, the need for a relevant policy framework for tourist SMEs in remote and peripheral areas should be stressed. This framework would enhance the potential of peripheral areas to follow the trend and adopt ICTs in support of their developmental strategic objectives.

References

Andrews, K. (1971), *The Concept of Corporate Strategy*, Homewood, Irwin.

Archdale, G. (1993), 'Computer Reservation Systems and Public Tourist Offices', *Tourism Management*, 14(1): 3–14.

Buhalis, D. (1997), 'Information Technology as a Strategic Tool for Economic, Social, Cultural and Environmental Benefits Enhancement of Tourism at Destination Regions', *Progress in Tourism and Hospitality Research*, 3(7): 71–93.

Buhalis, D. (1998), 'Strategic Use of Information Technologies in the Tourism Industry', *Tourism Management*, 19(5): 409–21.

Buhalis, D. (2003), *eTourism: Information Technology for Strategic Tourism Management*, Pearson Education Limited, England.

Buhalis, D. and W. Schertler (1999), *Information and Communication Technologies in Tourism*, Springer, Wien.

Chetty, S. and H. Wilson (2003), 'Collaborating with Competitors to Acquire Resources', *International Business Review*, 12: 61-81.

Cho, V. (1998), 'World Wide Web Resources', *Annals of Tourism Research*, 25(2): 518–21.

Connell, J. and P. Reynolds (1999), 'The Implications of Technological Developments on Tourist Information Centers', *Tourism Management*, 20: 501–509.

ENTER Program Report (1998), 'Technology and Change in Tourism', *Proceedings of the ENTER International Conference on Information and Communication Technology in Tourism*, Istanbul.

Fischer, D. (1998), 'Virtual Enterprise – Impact and Challenges for Destination Management', Paper presented at the *Conference on Information and Communication Technologies in Tourism*, ENTER 1998, 21-23 January, Istanbul.

Hudson, H. (1997), *Global Connections: International Telecommunications Infrastructure and Policy*, Wiley & Sons, New York.

Jagers, H.P.M., W. Jansen and G.C.A. Steenbakkers (1998), 'Wat zijn virtuele organisaties?: Op zoek naar Definities en Kenmerken', *Informatie en Informatiebeleid* (in Dutch), Spring No. 1: 61–7.

Kunz, R. (1998), 'Changes in World Tourism: From Marketplace to Marketspace', Paper presented at the *Conference on Information and Communication Technologies in Tourism*, ENTER 1998, 21-23 January, Istanbul.

Lafferty, G. and A. van Fossen (2001), 'Integrating the Tourism Industry: Problems and Strategies', *Tourism Management*, 22: 11–19.

Mutch, A. (1995), 'IT and Small Tourism Enterprises: A Case Study of Cottage–Letting Agencies', *Tourism Management*, 16(7): 533–39.

Naisbitt, J. (1995), *Global Paradox*, Avon, Aldershot.

Sheldon, P. (1997), *Tourism Information Technology*, CAB International, Wallingford.

Skyrme, D. (1998), 'The Realities of Virtuality', in P. Sieber and J. Griese (eds), *Organizational Virtualness, Proceedings of the VoNet – Workshop*, 27-28 April, Institute of Information Systems, Department of Information Management, University of Bern, pp. 25–34.

Stamboulis, Y. and P. Skayannis (2003), 'Innovation Strategies and Technology for Experience-Based Tourism', *Tourism Management*, 24: 35–43.

Swarbrooke, J. (1996), 'Technological Developments and the Future of the UK Tourism Industry', *Insights*, 7 (May): A173–A183.

Tacgmintzis, J. (1999), 'Information Society Technologies for the European Tourism Industry', Paper presented at the *International Conference on Information and Communication Technologies in Tourism*, ENTER 1999, 21-23 January, Innsbruck.

Talero, E. and P. Gaudette (1995), *Harnessing Information for Development: A Proposal for a World Bank Group Strategy*, The World Bank, <http://www.worldbank.org/html/fpd/telecoms/harnessing/>.

Tuntev, Z. (1998), 'Tourism and the Internet', Paper presented at the *First International Scientific Congress on Tourism and Culture for Sustainable Development*, 19-21 May, Athens.

Tsolakidis, C. (1998), 'Internet and Information Technology Means of Sustainability in Tourism', Paper presented at the *First International Scientific Congress on Tourism and Culture for Sustainable Development*, 19-21, Athens.

Wayne, S. (1998), 'Tourism 2020 Vision', Paper presented at the *Conference on Information and Communication Technologies in Tourism*, ENTER 1998, 21-23 January, Istanbul.

ENTER (1998), 'Enter Program Report', presented at the *Conference on Information and Communication Technologies in Tourism*, ENTER 1998, 21-23 January, Istanbul.

World Tourism Organization (WTO) (1998), 'Tourism 2020 Vision', Paper presented at the *Conference on Information and Communication Technologies in Tourism*, ENTER 1998, 21-23 January, Istanbul.

Chapter 7

European Informational Cultures and the Urbanization of the Mediterranean Coasts

Lila Leontidou

7.1 Introduction

European cultures are increasingly characterized by movement and hybridity due to population movements, among which tourism and migration play the major role. The impact of ICT expansion on movement has been profound, and movement, in turn, creates fluid local cultures. This web of interactions will be approached in this chapter from the vantage point of the cities, islands and coasts of Mediterranean Europe. It will be related to regional development indirectly, because our focus remains on the variety of cultures created by and spurring population movements, especially tourism, and their relationship with Mediterranean landscapes on the local rather than the regional scale.

On the methodological level, the 'cultural turn' in geography and the social sciences is relevant and indeed interesting for a theoretical approach to tourism. This in turn furnishes us with the epistemological and theoretical background for understanding those population movements which relate to new forms of tourism and especially new intra-European moves, rather than the usual questions posed by cultural studies of tourism which relate to the level of a representational analysis of 'gazing'. In fact, despite the relevance of the 'cultural turn' for issues discussed in this chapter, we consider the 'tourist gaze' type of approach as rather static. It offers little to the understanding of cultural tourism change and transformation, since it views cultural tourism as a type of discourse based on recycling and reconfirmation of memory, with an emphasis on landscape, discourse, or writing. But memory is constantly being reconstructed, not just reconfirmed; it changes as the modalities of the circulation of people, images and things change over time (Crang, 1998). The crystallization of discourses on memory and their commodification for global tourism flow from cultural and political imaging and positionality at both the global and the local level, interact with material reality, and affect cultural landscapes. What has been statically perceived in the cultural studies of tourism in the 1980s, is in fact a constantly changing process. The Mediterranean is a formidable milieu to make this point.

7.2 Spatial utopias in European cultures

Tourism is (or should be) the most escapist, liberated and spontaneous of activities outside work discipline and the Fordist regime. The practice of tourism as a *performative* activity and a *circulative* culture (Squire, 1994) makes people turn towards their *utopias*. They seek to enjoy a few days next to their imagined favourite landscapes, in activities which may range from creative searches to hedonistic consumerism. Let us begin with a distinction between the different constructions and positionalities by the different actors in tourism: the tourist, the tour operator, the host society each have different perceptions and priorities and construct interesting antitheses and distinctions. However, there are further shades of positionalities if we take the tourist as an example. The decision to become a tourist, to begin with, has been related to the hedonistic 'spirit of modern consumerism', focussed on pleasure rather than functionally interlinked needs. This has been stressed (and criticized) as a characteristic of the postmodern condition, since it clearly contradicts the taylorized work discipline (Albertsen, 1988; Leontidou, 1993: 958). But, in the course of its history, the Mediterranean has been flooded with travellers, who have also been explorers, pilgrims, and eco-tourists. Their motivation was exploration and pilgrimage rather than hedonistic consumerism, which also played a part, but was not the central concern. What are the pursuits and hopes of visitors to Athens, Dubrovnik, Sarajevo? Are they gravitating towards pleasure or education? Nostalgia or anticipation of the future? Are they seeking authenticity, staged authenticity, eclecticism, or landscapes immortalized by myth and historical accounts? Are they magnetized by reality or virtuality? Are visitors to the Balkans really seeking consumerism?

The beginnings of European tourism can be traced back to travellers, who were more like pilgrims to mythical places and antique treasures. They frequently stopped by in Southern Europe. A case study of Greece is interesting in order to observe the shift from philhellenes, intellectuals, poets, and of course the archaeologists, who still have a monopoly in excavations over their Greek colleagues, to mass tourism. After years of war, civil war, and reconstruction, tourists had different destinations: the sun and the beaches of the Greek islands concentrated most of the holiday makers, but the places of heritage in the islands and in the mainland were also visited (Leontidou, 1998). Among these places of heritage and their monuments, the Sacred Rock of the Acropolis is one of the most celebrated monuments in Europe although it has lost its full name and is now just 'the Acropolis'; but it is not a theme park. The Acropolis stands as one of the main spectacles representing the origins of European culture (Loukaki, 1999). European urban landscapes are frequently influenced by external civic models, especially those of Ancient Greece, and in the case of several cities, such external influence was overwhelming. 'The Acropolis' has been reproduced as an architectural style in different places throughout the world, and is itself still visited as a major monument. However, situated as it is in the heart of the untidy, playful, speculative, polluted and congested city of Athens, it has not saved the city from the major loss of tourist arrivals which hit it when negative publicity about environmental degradation took its toll over the last two decades. The Acropolis is now engulfed in unimaginative dense modernist structures, which ironically comprise an eclectic, postmodern collage of styles and land uses in a city which has resisted planning thoughout the last two centuries. This has been heavily criticized, and not only by the local urban planners. The tourists themselves have actively undermined speculative landscapes, by abandoning the city of Athens until the year 2004, when it hosted the Olympic Games.

This urban dystopia – definitely familiar in countries which underwent the industrial revolution – brings us to the main distinction made in this chapter. For a very long period until our own time, there has always been a contrast between the spatial expression of urban congestion and the rural idyll. Urbanism and anti-urbanism, as a relevant antithesis between Southern and Northern Europe, creates an interesting contrast of utopias envisaged in narratives lived and acted. Such utopias have been changing, inscribed within cultural traditions which go much deeper than personal preferences. The tradition of *urbanism in the Mediterranean*, which considers the city as the space of culture and virtue, contrasts with the *anti-urbanism of Anglo-American cultures*, which emerged especially after the Industrial Revolution. According to Schorske (1998), the latter created an aversion to cities of vice, criminality, poverty, congestion and pollution, reconstructed pastoral idylls, and ejected the rich to the suburbs. This pattern has been quite familiar to places which underwent the Fordist/post-Fordist transformation. On the other hand, Mediterranean urbanism emanates from certain diachronic characteristics of Mediterranean cultures which have to do with the cultural role of city states in Southern Europe (Leontidou, 1990, 2000). Mediterranean and Anglo-American conceptions of leisure, recreation, education, exploration and tourism, differ so much that they can be considered as two extremes in a continuum with many in-between spaces.

This North/South contrast has been studied in terms of its effects on urbanization and migration patterns (Leontidou, 1990, 2000), but hardly ever in terms of its influence on leisure and tourism, which is the focus of this chapter. Besides the tourist interest in major monuments of world heritage and leisure resorts, is it urban or rural spaces that attract tourists from various parts of Europe? And what kind of temporalities/seasonalities and spatialities/settlements do they create? Monuments may constitute the dominant narrative about local cultures at least until the last century (and millennium), but underneath there was the dimension of utopia in tourist waves: the urbanism/anti-urbanism antithesis.

This antithesis emerges in tourism waves with the Mediterranean as a destination and is clearly reflected in the statistics (see Leontidou, 1998). Tourists originating in different socio-spatial milieux, sought different aspects and enjoyed different facets of the cultures visited. In the present multicultural and multifunctional tourist waves – including urban tourism, sun-and-beach, alpine tourism and ecotourism – it is fascinating to observe the destinations of the various nationalities clustering in different localities of Greece, according to their own cultural stereotypes and, of course, the policy of travel agents and tour operators. The two poles of 'preferences' are held, on the one hand, by Americans and the Japanese, who gravitate towards urban tourism, and, on the other, by Europeans, who move as a mass and gravitate towards the islands, especially Rhodes and Northern Crete (Leontidou, 1991). Among Europeans, there are sharp differences between, for example, Spaniards, who tend to prefer Athens, and Scandinavians, who visit the Dodecanese islands en masse. In addition, while Athens has tended to become less attractive over time, the preferences of Spaniards, Italians and the French to that city correspond to positive representations of urbanism in the Mediterranean and to the urban-oriented lifestyles in Southern Europe, in contrast to more negative representations in Northern Europe (Leontidou, 1998). These create different urbanization trajectories, different city-building processes, different urban governance formations, and different tourism consumption patterns (Leontidou, 2000).

Though commodification and globalization do not overturn the basic trend of urbanism/anti-urbanism antitheses, these patterns have been more recently moderated and shifted towards urbanism, in the context of postmodernism and urban competition, which gentrify many urban cores and boost urban tourism in the entrepreneurial city (Craglia et al., 2004). Throughout the world, tourism is, no doubt, commodified. Relevant theory formulated in the Anglo-American environment, based on the establishment and the demise of Fordism, is *not* appropriate for Mediterranean societies, many of which skipped modernism and passed easily to the informational cultures of postmodernism, in the context of their informal economic cultures and spontaneous urbanization (Leontidou, 1993). This chapter attempts an alternative theorization based on such reversals of dominant grand narratives.

7.3 Littoralization with residential tourism

The cultural traditions of northerners against urbanism, combined with new opportunities for free movement of the population in Europe provided by European unification especially after the Maastricht Treaty (1992), have shifted populations to several new destinations within the EU. New realities after the fall of borders between East and West Europe, since the demolition of the Berlin Wall in 1989, created more intense migration to destinations which once used to be places of origin of economic migrants: Greece, Italy, Spain and Portugal became major destination places for various types of migration. Globalization and European integration have brought new waves of migrants, *flâneurs* and visitors from the whole of Europe – and the whole world – into the Mediterranean. Among them, tourist waves changed and new types of tourism emerged.

The Mediterranean region has always been under strong seasonal population pressures during the summer months, which created major concern. In the early 1990s, Mediterranean Europe received about 113 million package tourists a year (Perry and Ashton, 1994), and in 1990 it accounted for 34 per cent of all international tourists in the world (Lanquar, 1995). Several important recent changes, described below, are due to EC enlargement towards the Mediterranean in the 1980s, combined with ICT expansion and postmodern change in consumption patterns. These changes created cultural-economic restructuring trends which affected patterns of tourism and migration. The main offshoot was the expansion of what we have elsewhere called *residential tourism* (Leontidou and Marmaras, 2001), familiar from France but now spreading towards the west (Spain) and east (Greece) coasts of the Mediterranean. This type of migration has had an intense impact on the littoralization observed in Europe since the 1980s, but more intensively since the 1990s. Some findings of our case studies in the coasts and islands of the two aforementioned nation states will be presented at this point.

The concept of international residential tourism is rather new and not included in most books on tourism with few exceptions (Apostolopoulos et al., 2001). It is defined as the seasonal relocation of people to foreign resorts for long periods of time (several months rather than weeks), usually in owner-occupied houses. Their numbers have increased since 1993, because the Maastricht Treaty (1992) opened prospects for European citizens to settle and buy property in any Member State anywhere within the European Union (Leontidou, 1997). In addition, the Schengen Agreement removed other institutional barriers for movement among EU countries. The development of ICT and transport technology is an additional channel facilitating frequent movement between work and leisure, even across boundaries, as well as teleworking in the resorts where these people settle.

International residential tourism has initially been studied almost exclusively as a consumption-related phenomenon. The most sizeable and the best-studied category in research during the 1990s was the retirement migrants moving to the Mediterranean coasts and islands (Cribier, 1982; Barke, 1991; Oberg et al., 1993; Warnes, 1994; Williams et al., 1997; Buller et al., 1994). This type of migration has been distinguished from labour migration during the Fordist stage, in many respects: consumption-led migration predominates over production-led migration; movements from North to South reversed the movements of the 1960s and 1970s; and migration from cities to the countryside formed the core of the movements.

Economic non-activity reached as high as 67 per cent in Catalonia and 73 per cent in Languedoc among EU residents in 1990 (Lardies Bosque, 1997) and was not restricted to the elderly retirement tourists (Perry et al., 1986). Retirement tourists were encouraged by certain countries, because they smoothed out seasonal fluctuations: in Spain, legislation was passed to help them use their pensions locally (Leontidou and Marmaras, 2001). Consequently, tourism to Spain exploded, increasing tenfold in 1960–94, from 6 million in 1960 to over 61 million, of whom 88.8 per cent were European (SGT, 1995; EC, 1997; Lardies Bosque, 1997). This unprecedented increase was mostly composed of tourists who converted from seasonal visitors into residential tourists. Other Mediterranean destinations received less residential tourists, while Africa and the Middle East did not participate in this phenomenon to any significant degree.

However, although consumption-led migration predominates even at present, increasingly as a small part of these tourism/migration waves are composed of economically active populations, besides the pensioners or leisure migrants who were predominant during the beginning of the 1990s (Leontidou and Marmaras, 2001). This started with a tendency for investment in tourism-oriented businesses and facilities by residential tourists or returning migrants throughout Southern Europe, from Italy (King, 1984) to Portugal (Mendonsa, 1983). Emigrants returning from the North, or *retornados* from Africa, set up their businesses along the coast. These require little capital investment. In addition, small-scale economic activity increased in islands and tourist resorts among domestic residents and residential tourists: they used their property for income, where property included small workshops, besides using the house for themselves and their family. In a few cases, broader types of domestic tourist businesses have been studied (Williams et al., 1989). Later on in the 1990s, the number of economically active residential tourists rose and many international tourists and return migrants were actively involved in entrepreneurship and productive initiatives (Lardies Bosque, 1997). Small-scale enterprises were encouraged by policy makers, as large corporate involvement in the tourist industry came under increasing strain in Mediterranean Europe (Leontidou and Marmaras, 2001).

However, the decision to migrate to the coasts and tourist resorts has not been found to be related to these business activities. It is personal lifestyle reasons rather than economic considerations that have attracted foreign entrepreneurs to the Mediterranean coasts. Employment has followed the move, and has eased comfortable and more permanent settlement of residential tourists according to the survey in Spain (Lardies Bosque, 1997). It was found, for example, that the main attraction for entrepreneurs to move to the Costa del Sol was the pleasant climate (60 per cent) and the lower cost of living (Eaton, 1995; Lardies Bosque, 1997). Again, it is not economic decision making which motivates economic actors. Such findings underline the persistence of consumption-led migration in the phenomenon of residential tourism, contrasting with production-led migration during the Fordist phase until the 1960s. Naturally, this poses the problem of the viability of the residential tourists' enterprises.

In the geography of residential tourism, the urbanism/anti-urbanism antithesis flickers in destinations, but at the same time a coastal/inland contrast arises, as detailed in Leontidou and Marmaras (2001: 262–3):

> Each Southern European country has its own geography of foreigners. In France, rural areas are more attractive than coastal ones (Tuppen, 1991): a comparison between Spain and France found 64 per cent of migrants in Catalunya living in coastal municipalities,

contrasting with only 19 per cent of migrants in Languedoc. Another interesting aspect of residential patterns is the tendency of foreigners to conglomerate by nationalities in the various regions of Europe. In Languedoc, 65 per cent of non-French EU residents are Spanish (Lardies Bosque, 1997). In Italy there were about 40,000 Germans and 30,000 British citizens in 1991 (Leontidou, 1997). Many were scattered in the Sylvan landscapes, renovating abandoned houses of Tuscany and Umbria, to the disadvantage of locals wishing to buy property (King, 1991: 78; Pedrini, 1984).

The comparison between the coasts and the islands of Spain and Greece and the in-depth analysis of the causes and effects of residential tourism in selected localities of the two countries through institutional interviews and questionnaires to residential tourists, has revealed different profiles of foreigners in the two countries. Residential tourists attracted to Spain are part of a mass-tourism process linked with retirement tourism and organized by agents in the home countries. Those choosing Greece, by contrast, move individually and act through networks of friends rather than through organized agents (Marmaras, 1996). Many seek isolation and alternative lifestyles. On the basis of questionnaires posted to 14 per cent of foreigners in the Greek islands and the case study of Spain, the profile of the international residential tourists in Spain and Greece can be compared:

• Retirement tourists over 65 years old were a minority (16 per cent) in Greece, while they were 25 per cent of the total in Spain.
• The resorts filled with older residents in the winter and younger ones in the summer, in a sort of intra-family time-sharing in Spain, especially in the high-rise buildings of Benidorm and Torremolinos which create controversial landscapes nonexistent in Greece.
• Most residents had a university degree (72 per cent) in Greece, compared with only 45 per cent in Spain.
• Greek residential tourists originated from Britain (38 per cent), Germany (19 per cent), the Netherlands (14 per cent; Marmaras, 1996), but Spanish residential tourists came mostly from Britain.
• Gender differences were not very important, but foreign women have tended to predominate among the residential tourists who came to stay in Greece after marrying a native: 25 per cent of foreigners in Greek islands fell in that category.

Residential tourism waves rest upon culturally diverse representations, consumption and leisure patterns in the societies of origin. In the societies of destination, these come to be differently embedded: they fall into the localities concerned, which respond in different ways, construct different cultural stereotypes, and build different landscapes to host such migrants. The resident/tourist interaction thus presents a diversity of articulations in different localities, and certain stereotypes prevalent until the 1980s must be now dismissed. The 'erosion' approach considers a static landscape wherein the local society is eroded by sun-and-beach tourism and its secular values. According to these researchers, local societies are inert. However, in our comparison of Mediterranean coastal areas, we found local societies in flux, the re-invention of tradition by global agents and a local population reconstructing ways of life in spite of these redefinitions, recycling memories, constructing landscapes. We found a geography of events, happenings, occurrences, a space of flows and of becomings (Crang, 1998). We also found a new population coming from several worlds, changing in time; real and virtual worlds, local/'authentic', or global, eclectically incorporating voices from other cultures and traditions. Change and eclecticism constitute a departure from 'authenticity' as well as a departure from the view of landscape as text. Postmodern Europe is reflected on the Mediterranean coasts.

7.4 Virtual reality and informational cultures

Residential tourism is facilitated by ICT and the possibilities these technologies open for teleworking and moving between places in Europe, as already pointed out above. On a more general level, the digital revolution and the emergent information society have boosted several hi-tech fantasies, and what is already happening in the realm of tourism has many affinities with these imaginings. What itineraries does a prospective tourist construct in a living room with literature, music, TV newsreels and documentaries, or in a cinema or a concert hall at home? To what extent is everyone attracted by representations, by simulacra, or by material objects such as souvenirs? Is the prospective tourist influenced by advertisements or the promotion emanating from tour operators and urban managers, or by virtual accounts and recommendations by friends?

It is worth introducing here yet another consequence of the informational age which relates to the Mediterranean (Craglia et al., 2004): tourism generated by best-sellers, movies and TV spectacles, even the most negative ones of war and destruction. Balkan experiences stir all types of contradictions, with the striking contrast between cultural/heritage/entrepreneurial tourism in Athens, on the one hand, and war tourism in Sarajevo and Mostar, with Dubrovnik as an in-between space, on the other. Events-oriented tourism is illustrated in the rising numbers of visitors to cultural capitals, to Olympic cities, but also to war-stricken areas. As war became a TV spectacle in the Balkans during the 1990s, war was commodified. The best-known and most celebrated example is the inscription in front of the Dubrovnik gate, detailing the plan of the city with reference to the buildings destroyed, and leading the tourists to sites which were not renovated, so as to serve as spectacles of war. In Mostar at the turn of the millennium, the tourist guides were as detailed on war and destruction as the devastated cityscape, with its bullet-damaged buildings and demolished bridges. Such diverse and clashing experiences and types of tourism, as found in the Balkans today, undermine the relevance of general theory and 'grand narratives'. They also introduce the rapid change and transformation of *performative* tourism at the turn of the millennium, today.

Virtual realities and the media lead to a construction of hybrid spaces of leisure, corresponding to cultural imaginings of each and every visitor, according to their readings, viewings, spatialities, utopias and dystopias, and geographical imaginations (Gregory, 1994) in general. North/South antitheses in spatialities are diluted in hybrid spaces on the cultural level and reshuffle the practice of tourism as a performative activity and a circulative culture (Squire, 1994). Tourism in the new millennium turns the discussion to radically different paradigms and epistemologies, but, more important, to different ontological questions than simple hedonistic consumerism. Eclecticism, hybridity and the polyvalence of tourist flows come to the fore. If Anglo-American rock groups receive standing ovations anywhere in the world, if ethnic restaurants are the centre of entertainment in London or Paris, nevertheless, who would expect to encounter a celebration of Tolkien within the traditional castle of Gorizia, complete with costumed entertainers, memorabilia and artwork, in 1997, long before the trilogy was filmed? Cities host concerts, exhibitions and events from remote places in a world of virtual reality, flows and interaction, and no longer consider this as an 'erosion' of local cultures.

On the economic level, ICT expansion creates new opportunities of distance learning, teleworking and communication in general, which is expected to affect population densities in the sunbelt zones of Europe. Trends are already evident in our case studies of Spain and Greece. International residential tourism tends to surpass the 'leisure' category and acquires the character of a postmodern lifestyle, where spatial fixity in one home is in question (Leontidou, 1993). Studies of elderly migration in North America have found that seasonal residence in warm locations has become a way of life. This trend is now becoming international, and spreading to Europe as a whole. The term 'international counter-urbanization' (Perry et al., 1986; Buller and Hoggart, 1994) indicates a movement of population and real estate activity from the larger Northern cities to the Southern coasts and islands. The settlement network is affected by southward population shifts and gravitation toward the coast. Processes of 'littoralization' are presently at work (Leontidou et al., 1998). It is difficult to project them in the future, given global warming; however, the present is full of reversals.

7.5 Conclusion

The most interesting aspects of new migration trends in Europe today, relate to several reversals observed along the Mediterranean coasts and throughout Europe, on most spatial levels: urban/rural, North/South, East/West, production/ consumption-led migration, and others. There are several novel trends here, for both Southern and Northern Europe. The passage from emigration to immigration is the most important aspect for the South (King and Rybaczuk, 1993), whether in the case of labour migrants from the Third World, repatriation, or residential tourists from the rest of Europe. The economic migration trend, which has been characterized as Fordist in the past (Fielding, 1993), subsided in Europe by the 1970s (Leontidou, 1990; King, 1993a, 1993b). There was a reversal from production to consumption-led migration. The counter-urbanization of the whole of Europe extends over most seasons and is no longer a summertime phenomenon. It reverses the settlement trends as well as seasonality and yet other phenomena of the past:

> In postmodern Europe, the increase of seasonal residential mobility reflects material affluence and changing lifestyle preferences, combined with the opening of European borders and the improvement of communication technology. This new type of *seasonal/ semi-permanent residential mobility* spread from pensioners, to more people seeking a more varied life style, where autumn and winter, spring and summer, are spatially as well as temporally differentiated. The corresponding adaptation of the job market to follow this trend, may be just a matter of time, as we are moving to the new millenium (Leontidou and Marmaras, 2001: 267).

Mediterranean urban tourism has never been Fordist: even mass tourism was embedded in informality in the Balkans and Southern Europe. The hidden economy has been dominant and informality has been prevalent in work commitment and accommodation, ever since the time when tourism first emerged. Polyvalence, seasonality and informality encountered in other spheres of Mediterranean life, have been much more pronounced in tourism than in manufacturing industry or in housing and land allocation. In other words, the combination of informality and the information society should not be conceptualized as post-Fordism here. Mediterranean tourism does not fit into the transition posited by regulation theory – from Fordism to post-Fordism. The former has hardly ever materialized in Southern Europe and the Balkans. If we strive for systematization, we could consider theories of modernism/postmodernism, or of orientalism and its negation, or of informal deregulation.

Tourism evades official norms and regimes of regulation. It has been always un-Fordist, even anti-Fordist rather than post-Fordist. It is increasingly commodified and globalized, but also hybridized as its forms change and become polyvalent. This is even more so in the Mediterranean, where several activities which have developed in an environment of spontaneity and trespassing – informally, unpredictably – are now drawn into competition and entrepreneurialism. Still, however, tourism cultures include the coexistence of opposites. They deconstruct dichotomies with in-between spaces: the global and the local is glocalized; the sacred and the secular may coexist; embeddedness and disembeddedness are differentiated; entrepreneurialism coexists with the *genius loci*. Representations of tourist landscapes have been constructed globally but remain local, as the case of Athens amply demonstrates. The 'proprietors' of ancient monuments in history were not necessarily today's local residents, but the latter actually manage and savour the monuments' aura. The same goes for the Greek islands, but in this case the trend is towards increasing marketing to Europeans other than Greeks. Differences of leisure patterns and of destinations according to the origin of tourists, undermine general theory and modernist 'grand narratives' in tourism research, despite the fact that the defining features of tourism regimes have to account for the globalization aspect, which is cultural in nature. Now, as global tourism continues to rise at the turn of the millennium, local narratives and uniqueness of place must be defended, without necessarily leading to continuous theoretical fragmentation. They can be defended from a theoretical and an ontological point of view, in the context of postmodernism.

References

Albertsen, N. (1988), 'Postmodernism, Postfordism and Critical Social Theory', *Environment and Planning D: Society and Space*, 6: 339-365.

Apostolopoulos, Y., S. Leivadi and A. Yiannakis (eds) (1996), *The Sociology of Tourism*, Routledge, London.

Apostolopoulos, Y., P. Loukissas and L. Leontidou (eds) (2001), *Mediterranean Tourism: Facets of Socio-economic Development and Cultural Change*, Routledge, London.

Barke, M. (1991), 'The Growth and Changing Pattern of Second Homes in Spain in the 1970s', *Scottish Geographical Magazine*, 107(1): 12–21.

Blacksell, M. and A. Williams (eds) (1994), *The European Challenge: Geography and Development in the European Community*, Oxford University Press, Oxford.

Buechler, H.C. and J.M. Buechler (eds) (1987), *Migrants in Europe*, Greenwood, Westport.

Buller, H. and K. Hoggart (1994), *International Counterurbanization: British Migrants in Rural France*, Avebury, Aldershot.

Craglia, M., L. Leontidou, G. Nuvolati and J. Schweikart (2004), 'Towards the Development of Quality of Life Indicators in the "Digital" City', *Environment & Planning B: Planning and Design*, 31(1): 51-64.

Crang, M. (1998), *Cultural Geography*, Routledge, London

Cribier, F. (1982), 'Aspects of Retirement Migration from Paris: An Essay in Social and Cultural Geography', in A.M. Warnes (ed.), *Geographical Perspectives on the Elderly*, Wiley, London, pp. 111–37.

Dunford, M. and G. Kafkalas (eds) (1992), *Cities and Regions in the New Europe*, Belhaven Press, London.

Eaton, M. (1995), 'British Expatriate Service Provision in Spain's Costa del Sol', *The Service Industries Journal*, 15(2): 251–66.

Fielding, A.J. (1993), 'Mass Migration and Economic Restructuring', in R.L. King (ed.), *Mass Migrations in Europe: The Legacy and the Future*, Belhaven Press, London, pp. 7–18.

Gregory, D. (1994), *Geographical Imaginations*, Blackwell, Oxford.

King, R.L. (1984), 'Population Mobility: Emigration, Return Migration and Internal Migration', in A. Williams (ed.), *Southern Europe Transformed*, Harper & Row, London, pp. 145–78.

King, R.L. (1991), 'Italy: Multi-faceted Tourism', in A. Williams and G. Shaw (eds), *Tourism and Economic Development: Western European Experience*, 2nd edn, Wiley, London, pp. 61–83.

King, R.L. (ed.) (1993a), *Mass Migrations in Europe: The Legacy and the Future*, Belhaven Press, London.

King, R.L. (ed.) (1993b), *The New Geography of European Migrations*, Belhaven Press, London.

King, R.L., P. De Mas and J.M. Beck (2000, eds) *Geography, Environment and Development in the Mediterranean*, Sussex Academic Press, Brighton.

King, R.L., L. Proudfoot and B. Smith (eds) (1997), *The Mediterranean: Environment and Society*, Edward Arnold, London.

King, R.L. and K. Rybaczuk (1993), 'Southern Europe and the International Division of Labour: From Emigration to Immigration', in R.L. King (ed.), *The New Geography of European Migrations*, Belhaven Press, London, pp. 175–206.

Kinnaird, V. and D. Hall (eds) (1994), *Tourism: A Gender Analysis*, John Wiley & Sons, New York.

Lanquar, R. (1995), *Tourisme et Environnement en Mediterrannee*, Les Fascicules du Plan Bleu 8, Economica, Paris.

Lardies Bosque, R. (1997), 'Migration and Tourism Entrepreneurship: North-European Immigrants in Mediterranean Coasts', Unpublished Fellow's Report, in L. Leontidou (coord., 1997), *Migration and Tourism Development in Marginal Mediterranean Areas*, Unpublished Fellows' Reports for the EU DG XII Human Capital and Mobility Project, King's College London.

Leontidou, L. (1990), *The Mediterranean City in Transition: Social Change and Urban Development*, Cambridge University Press, Cambridge.

Leontidou, L. (1991), 'Greece: Prospects and Contradictions of Tourism in the 1980s', in A. Williams and G. Shaw (eds), *Tourism and Economic Development: Western European Experience*, 2nd edn, Wiley, London, pp. 84–106.

Leontidou, L. (1993), 'Postmodernism and the City: Mediterranean Versions', *Urban Studies*, 30(6): 949-965.

Leontidou, L. (1994), 'Gender Dimensions of Tourism in Greece: Employment, Sub-cultures and Restructuring', in V. Kinnaird and D. Hall (eds), *Tourism: A Gender Analysis*, John Wiley & Sons, New York, pp. 74–105.

Leontidou, L. (1995), 'Repolarization in the Mediterranean: Spanish and Greek Cities in Neoliberal Europe', *European Planning Studies*, 3(2): 155–72.

Leontidou, L. (1997) (coord.), *Migration and Tourism Development in Marginal Mediterranean Areas*, Unpublished Fellows' Reports (E. Fernandez Martinez, R. Lardies Bosque, E. Marmaras) for the EU DG XII Human Capital and Mobility Project, King's College London.

Leontidou L. (1998), 'Greece: Hesitant Policy and Uneven Tourist Development in the 1990s', in A. Williams and G. Shaw (eds), *Tourism and Economic Development: European Experiences*, 3rd revised edn, Wiley, London, pp. 101–23.

Leontidou, L. (2000), 'Cultural Representations of Urbanism and Experiences of Urbanisation in Mediterranean Europe', in R.L. King, P. De Mas and J.M. Beck (eds), *Geography, Environment and Development in the Mediterranean*, Sussex Academic Press, Brighton, pp. 83–98.

Leontidou, L., M.L. Gentileschi, A. Aru and G. Pungetti (1998), 'Urban Expansion and Littoralisation', in P. Mairota, J.B. Thornes and N. Geeson (eds), *Atlas of Mediterranean Environments in Europe: The Desertification Context*, John Wiley & Sons, London, pp. 92–7.

Leontidou, L. and E. Marmaras (2001), 'From Tourists to Migrants: International Residential Tourism and the 'Littoralization' of Europe', in Y. Apostolopoulos, P. Loukissas and L. Leontidou (eds), *Mediterranean Tourism: Facets of Socio-economic Development and Cultural Change*, Routledge, London, pp. 257–267.

Mairota, P., J.B. Thornes and N. Geeson (eds) (1998), *Atlas of Mediterranean Environments in Europe: The Desertification Context*, John Wiley & Sons, London.

Marmaras, E. (1996), 'Migration and Tourism Development in Marginal Mediterranean Areas: Foreign Second-home Owners in Spain and Greece', in L. Leontidou (coord., 1997), *Migration and Tourism Development in Marginal Mediterranean Areas*, Unpublished Fellows' Report for the EU DG XII Human Capital and Mobility Project, King's College London.

Mendonsa, E.L. (1983), 'Search for Security, Migration, Modernisation and Stratification in Nazare, Portugal', *International Migration Review*, 6: 635–45.

Oberg, S., S. Scheele and G. Sundstrom (1993), 'Migration Among the Elderly: The Stockholm Case', *Espaces, Populations, Societes*, 3: 503–14.

Pedrini, L. (1984), 'The Geography of Tourism and Leisure in Italy', *Geojournal*, 9: 55–7.

Perry, A. and S. Ashton (1994), 'Recent Developments in the UK's Outbound Package Tourism Market', *Geography*, 79: 313–21.

Perry, R., K. Dean and B. Brown (1986), *Counterurbanisation*, Geo Books, Norwich.

SGT (1995), *Nota de Coyuntura Turistica*, Enero, Secretaria General de Turismo, Madrid.

Squire, S. (1994), 'The Cultural Values of Literary Tourism', *Annals of Tourism Research*, 21(1): 103–20.

Sullivan, M. (1985), 'The Ties That Bind', *Research on Ageing*, 7: 235–60.

Tuppen, J. (1991), 'France: The Changing Character of a Key Industry', in A.M. Williams and G. Shaw (eds), *Tourism and Economic Development: Western European Experiences*, 2nd edn, Wiley, London, pp. 191–206.

Warnes, A.M. (ed.) (1982), *Geographical Perspectives on the Elderly*, Wiley, London.

Warnes, A.M. (1994), 'Permanent and Seasonal International Retirement Migration: The Prospects for Europe', *Netherlands Geographical Studies*, 173: 69–81.

Williams, A.M. (ed.) (1984), *Southern Europe Transformed*, Harper and Row, London.

Williams, A.M. (1997), 'Tourism and Uneven Development in the Mediterranean', in R.L. King, L. Proudfoot and B. Smith (eds), *The Mediterranean: Environment and Society*, Edward Arnold, London, pp. 208–26.

Williams, A.M., R. King and T. Warnes (1997), 'A Place in the Sun: International Retirement Migration from Northern to Southern Europe', *European Urban and Regional Studies*, 4(2): 115–34.

Williams, A.M. and G. Shaw (eds) (1991), *Tourism and Economic Development: Western European Experiences*, 2nd edn, Wiley, London.

Williams, A.M. and G. Shaw (eds) (1998), *Tourism and Economic Development: European Experiences*, 3rd revised edn, Wiley, London.

Williams, A.M., G. Shaw and J. Greenwood (1989), 'From Tourist to Tourism Entrepreneur, From Consumption to Production: Evidence from Cornwall, England', *Environment and Planning A*, 21: 1639–53.

PART II
Methodological Advances in Tourism Research

Chapter 8

Economic Impacts of Tourism: A Meta-analytic Comparison of Regional Output Multipliers

Eveline S. van Leeuwen, Peter Nijkamp and Piet Rietveld

8.1 Introduction

Tourism history as it appears in much (Western) literature is largely concerned with the activities of the affluent, which occur in particular tourism settings such as resorts or on luxury tours. Originally, these forms of tourism were special events, which occurred periodically in people's lives. It all began with the leisured elites of ancient Greece and Rome, the re-emergence of tourism in the Renaissance, and the development of spas and Grand Tours in the seventeenth and eighteenth centuries. In the nineteenth century, tourism was heavily affected by the advent of the train and in the twentieth century by the jet aeroplane and charter flights.

In this process, tourism is seen as dispersing geographically increasingly outwards from the heartlands in Western Europe, creating a series of 'pleasure peripheries', and spreading socially from the upper classes, down through the middle ranks and ultimately to the mass working classes (Towner, 1995). Tourism became a 'normal' economic good in a welfare society.

Nowadays, we define tourism as 'the activities of persons travelling to and staying in places outside their usual environment for not more than one consecutive year for leisure, business and other purposes' (UN/WTO, 1994: 5). People often speak of the impact of the tourism sector, but one should be aware of the fact that no such thing exists as '*the* tourism sector' in any usual statistical definition. Even a sector that is strongly oriented towards leisure activities, such as the hotel and restaurant sector, also serves non-leisure business activities (Oosterhaven and van der Knijff, 1988).

In many regions (both urban and rural), tourism is increasingly seen as a means for economic development. Foreign tourists spend money in tourist areas, and in this way a country earns foreign currency which leads to an improvement of the economic situation of the country concerned. Less prosperous regions, which have an attractive environment or cultural heritage, could be developed as tourist destinations to stimulate the regional economy (Baaijens et al., 1997). This has recently prompted much policy and research interest in the benefits of tourism for regional income and employment. The development of tourism as a means to widen the export base and generate more employment is of primary importance. Policy makers in the government need to know the magnitude of the impact of international and domestic tourist expenditures on the economy in order to make decisions about budget allocations for the development of tourist facilities (Freeman and Sultan, 1999). But there is much variation, and the question emerges whether such variations can be ascribed to systematic factors.

In this chapter, we perform a meta-analysis on tourism multipliers. As multiplier values reflect the size of the multiplier effect, with respect to a specific feature of the economy such as income or employment, these values can help policy makers to learn something about the magnitude of tourist expenditures. Within a meta-analysis, the empirical outcomes of studies with similar research questions are analyzed (Baaijens et al., 1997). The research question we want to answer in this chapter is: Which characteristics of the tourism sector, the research area, or the type of publication in which a study appeared can explain variations in the size of the tourism multiplier?

Therefore, we first describe an economic model, viz. the input-output (I-O) model, which shows the flows of products or services and their interactions between several sectors (8.2). Next we explain the order of magnitude of the multipliers, followed by a first analysis of the empirical data (8.3). After this, we perform a linear regression on our available data (8.4). Another meta-analytical method, viz. Rough Set Analysis, will be used in the next section (8.5). Finally, some research conclusions will be drawn (8.6).

8.2 Input-output analysis

The use of I-O models in estimating economic impacts of recreation and tourism has increased considerably in the past decades because of both its ability to provide accurate and detailed information and the ease of interpreting the results (Fletcher, 1989; Pindyck and Rubinfeld, 1991).

The basic Leontief I-O model is generally constructed from observed economic data for a specific geographic region (nation, state, county, etc.). The basic information dealt with in I-O analysis concerns the flows of products from each industrial sector considered as a producer to each sector considered as a user (Miller and Blair, 1985).

The great advantage of I-O models is their internal consistency. All effects of any given change in final demand can be recorded. Important assumptions made in the I-O model are that all firms in a given industry employ the same production technology (usually assumed to be the national average for that industry), and produce identical products. It is also assumed that there are no economies or diseconomies of scale in production or factor substitution. I-O models are essentially linear: doubling the level of tourism activity/production will double the inputs, the number of jobs, etc. This reveals something of the inflexibility of the model (Stynes, 1997). Thus, the model is entirely demand-driven but bottlenecks in the support of inputs are totally ignored. Still input-output analysis is seen as a very clear and important method, which has its limitations, but, on the other hand, is often embedded in new and flexible methods.

8.3 Multipliers

I-O models are constructed primarily because they provide a detailed industry-by-industry breakdown of the predicted effects of changes in demand. It is sometimes useful, however, to provide a summary statement of these predictions (Armstrong and Taylor, 2000). This can be done by constructing multipliers based on the estimated re-circulation of spending within the region; recipients use some of their income for consumption spending, which then results in further income and employment (Frechtling, 1994).

The three most frequently used types of multipliers are those that estimate the effects on: 1) outputs of the sectors; 2) income earned by households because of new outputs; and 3) employment expected to be generated because of the new outputs.

This generated effect appears at three levels. First, the *direct effect* of production changes. For example, an increase in tourists staying in a hotel will directly increase the output of the hotel sector. *Indirect effects* result from various rounds of re-spending of, for example, tourism receipts in linked industries. If more hotel rooms are rented, then more breakfast products or cleaning services are needed. This will have an indirect effect on these sectors. The third level of effects is the *induced effects*. These effects only occur in a closed I-O model because of changes in economic activity resulting from household spending of income earned directly or indirectly as a result of, for example, tourism spending. These households can be hotel employees, who spend their income in the local region (Stynes, 1997). When comparing the sizes of the multipliers, it is very important to distinguish the different effects, which are taken into account.

The size of the multiplier depends on several factors. First of all, it depends on the overall size and economic diversity of the region's economy. Regions with large, diversified economies which produce many goods and services will have high multipliers, as households and business can find most of the goods and services they need in their own region. In addition, the geographic scale of the region and its role within the broader region also plays a role. Regions of a large geographic coverage will have higher multipliers, compared with similar small areas, as transportation costs will tend to inhibit imports (imports are seen as leakage and have a negative effect on a multiplier). Regions that serve as central places for the surrounding area will also have higher multipliers than more isolated areas. Furthermore, the nature of the specific sectors being considered can have a significant effect. Multipliers vary across different sectors of the economy according to the mix of labour and other inputs and the tendency of each sector to buy goods and services from within the region (less leakage to other regions). Tourism-related businesses tend to be labour-intensive. They, therefore, often have larger induced effects, because of household spending, rather than indirect effects. Finally, the year of the development of the I-O table should be taken into account. A multiplier represents the characteristics of the economy at a single point in time. Multipliers for a given region may change over time in response to changes in the economic structure as well as to price changes (Stynes, 1998).

In the meta-analysis undertaken in this study, we look at output multipliers. The reason for this is that, in our sample of studies we found that these multipliers are most often deployed.

8.4 Data analysis of tourist multipliers

For our meta-analysis, we were able to collect 24 case studies, which contain estimates of tourist multipliers including an output multiplier. A precondition was that the multiplier had to be derived with help of I-O analysis. Also a (brief) description of background factors concerning, for example, the area and the tourist activities had to be given. Appendix I shows the characteristics we used together with the classification.

The database

More than half of the case studies are (non-refereed) reports found on the Internet. These reports are often written by researchers to give local authorities insight into the tourist situation of the area concerned. In addition, several articles from refereed journals have been included.

A quarter of the case studies collected are written as papers for scientific conferences. As Figure 8.1 shows, these papers report, on average, the highest multipliers, whereas the reports contain multipliers with relatively low values. We also incorporate the year in which data was gathered to build the input-output table. In two instances this was before 1990, and in nine instances it took place in 1999 or later. On average, Figure 8.1 shows that the oldest multipliers have the highest value and the newest ones the lowest.

Of course, not only characteristics of the reports affect the size of the output multiplier but also the characteristics of the areas which are described. More than half of the areas have a population density below 100 persons per square kilometre. As much as a quarter of the studies have a population density of less than 15 persons. These areas are mostly the places with high nature values. Those studies referring to areas with the highest population density are related to urbanized areas or cities.

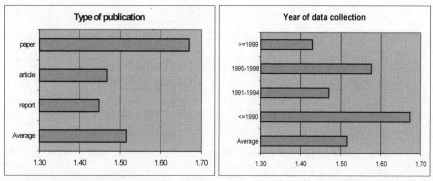

Figure 8.1 Average multipliers according to type of publication and years of data gathering

Figure 8.2 shows that multipliers concerning countries (type of area) are higher on average than, for example, the average multiplier values of a city. This is partly due to the fact that a city has to import a large part of its inputs, which leads to leakages. This also applies to a region. In fact, according to our database, regions have even lower output multipliers than cities.

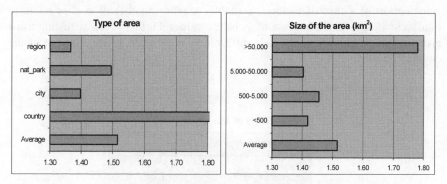

Figure 8.2 Average multipliers according to type of area and size of area

Figure 8.2 confirms this: the areas which are classified as regions often have a size of between 5,000 and 50,000sq km, and in this figure the class concerned shows a low average multiplier value. The largest areas also are related to large multipliers.

Of course, tourism itself also has to be included in this multiplier analysis. The next two graphs (Figure 8.3) show the average multiplier values related to the total expenditures as well as related to the number of visitors (:1000).

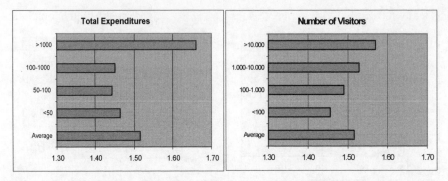

Figure 8.3   Average multiplier values according to total expenditures per year and number of visitors per year

According to Figure 8.3, higher output multipliers occur when more tourists visit an area and when they spend a lot of money. Of course, we have to keep in mind that multipliers themselves describe the effect of expenditures. But the total amount of expenditures and the total number of visitors also explain some of the characteristics of the tourism sector concerned.

Finally, it is interesting to know more about the travel motives of tourists. In most of the studies, the visitors are attracted to their tourist destination by a combination or a mix of factors. A slightly smaller number of studies relate to tourists visiting a region because of its nature values. Furthermore, we distinguish a group of studies in which people visit a place, most often a city, because of the cultural values.

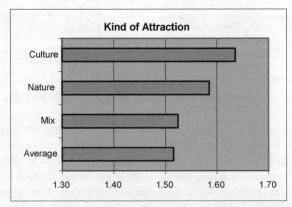

Figure 8.4   Average multiplier values concerning kind of attraction

According to Figure 8.4, this last group is related to the highest output multipliers. An explanation can be that people who visit areas with high cultural values, such as a city, often stay only a short time but (have to) spend a lot of money on plane tickets, food and often on hotels.

All these figures only show average multiplier values, but they do not reveal anything concerning any relationship or coherence between the different indicators and the height of the output multipliers.

8.5 Linear regression results

8.5.1 Introduction

Linear regression is a standard technique that can also be used in meta-analytical experiments, insofar as statistical results from a sample of previous studies are analyzed. This statistical model presupposes that there is a one-way causation between the dependent variable Y and the independent variable X. Because we have only a limited number of studies available, it is not plausible that the assumptions of the standard linear regression are satisfied, for the variances of the distance multiplier values are not equal (see also Baaijens and Nijkamp, 1996).

The analysis starts with the formulation of a set of hypotheses, which we will verify with the help of the linear regression model.

- *The larger the economic base, the higher the multiplier.* As described in the section on multipliers, a larger economy needs less imports. Imports can be seen as a leakage of money to other regions. Therefore, it may be expected that the size of land area or the size of the population has a positive effect on the multiplier. In particular, countries as a whole will have lower imports.
- *The more visitors or expenditures of visitors, the higher the multiplier.* The visitors or at least the expenditures of the visitors may cause a higher multiplier as a result of cluster effects.
- *The longer the time span for which the multiplier was derived, the higher the multiplier.* If we assume that the tourism sector changed during the years and became more internationally-oriented, the 'older' multipliers should be higher.

8.5.2 Results of the Regression Analysis

The estimation results for the output multiplier equations can be found in Table 8.1. When looking at the correlation between the variables with help of bi-variate Pearson correlation, we find that several variables are related. We find, for example, a positive significant correlation between density and culture, or city. Therefore some of these variables are excluded from the regression analysis.

The first equation focuses on the meta-variables. If we take into account R^2, which describes the proportion of the total variation in the dependent variable (the output multiplier) explained by the regression of the variables, we see that the meta-variables describe only 38 per cent of the variation. The equation shows us that multipliers published in an article are lower than those published in a conference paper. Furthermore the year of data collection is of importance, the newer the data, the lower the multiplier.

The second equation uses the variables related to the characteristics of the area concerned. It appears that especially the country dummy is significant. This dummy variable has the value 1 if the area is a country and 0 if the area is, for example, a county or a city. Because the country dummy shows a positive coefficient, this can indicate that the boundaries along countries, to a certain extent, prevent leakages.

Table 8.1   Regression equations of the output multiplier

Variable	Meta-variables	Area specific	Tourism specific	All	Selection
	1	2	3	4	5
Constant	1.973***	1.403***	1.480***	1.987***	1.911***
t-value	10.016	35.358	8.738	10.457	14.013
Year of data (0= 1980)	-0.029***		-0.019**	-0.026**	-0.028***
t-value	-2.694		-2.251	-3.448	-3.753
Conference Paper (dummy)	0.191*			-0.017	
t-value	2.018			-0.226	
Article(dummy)	-0.253*			-0.499***	-0.515***
t-value	-1.814			-3.229	-3.453
Density (100 inh/km²)		0.001		0.014***	0.011**
t-value		0.447		3.546	2.665
Population (1E 05)		0,001		0.003**	0.001**
t-value		1.360		2.630	1.909
Country (dummy)		0.296***		0.162*	0.198**
t-value		2.788		1.755	2.320
Nature (dummy)			0.349**	-0.070	
t-value			2.909	-0.669	
Mix (dummy)			0.268**	-0.164	
t-value			2.200	-1.587	
Visitors (1E 05)			0.002	-0.001	-0.0005
t-value			0.967	0.034	-0.302
R^2	0.38	0.64	0.43	0.90	0.81
N	24	22	24	22	22

Notes:
*** Correlation is significant at the 0.01 level (2-tailed)
** Correlation is significant at the 0.05 level (2-tailed)
* Correlation is significant at the 0.10 level (2-tailed)

When looking at the next column, with the tourism specific variables, we find again that the year of data is of importance as well as the nature and the mix dummy. These dummies describe the factor which attracts the visitors, e.g. nature values, cultural values, or a mix of characteristics. However, the effect of the visitors does not seem significant in any of the equations.

The fourth equation includes all variables distinguished for the regression analysis. We find five variables that show significant coefficients. As can be found in the table, the year of data has a negative effect on the multipliers. This means that, when the data are younger, the multipliers get lower. It also appears that when a multiplier is published in an article in a scientific journal, the value will be lower than when it is published in a paper or report. According to the values of the parameters, we may say that the output multiplier published in an article is 0.50 lower than the multiplier found in an average report commissioned by a client. Furthermore, the population density, the area and the country variable show positive coefficients with the output multiplier.

The fifth, and last, equation includes a selection of variables which were more or less significant in the previous equations. Again the year of data and the article dummy give negative coefficients and the population density, total population and country variables positive coefficients.

Looking back at the three hypotheses, we can conclude that the first hypothesis can be accepted: the larger the economic base, the higher the multiplier, and the effect of the country dummy as well as that of the population variable is positive and significant. Unfortunately, we can not agree with the second hypothesis: the number of visitors does not show a correlation with the outcome multiplier. But we can agree with the third one: the older the data on which the multiplier was derived, the higher the multiplier.

8.6 Rough set analysis

8.6.1 Introduction

Rough set analysis was developed in the early 1980s by Pawlak (1991). The method generally serves to pinpoint regularities in classified data, to identify the relative importance of some specific data attributes and to eliminate less relevant ones, and to discover possible cause-effect relationships by logical deterministic inference rules (van den Bergh et al., 1997). It is essentially a classification method devised for non-stochastic information. This means that ordinal or categorical information (qualitative data) can be taken into consideration. Quantitative data must first be converted into qualitative or categorical data by means of a codification (Nijkamp and Pepping, 1998).

The difference between the outcomes of linear regression and the rough set analysis is that the first of these methods indicates a potential causal relationship between the dependent and the independent variable. The relationship between a decision attribute (the dependent variable) and the condition attribute (the independent variable) of the rough set analysis refers to the statistical frequency at which a certain category of the decision attribute occurs in certain categories of the condition attributes (Oltmer, 2003). The classification of the values of the attributes is a somewhat problematic issue in the application, as the use of thresholds implies some loss of information, and the thresholds are chosen subjectively (Bruinsma et al., 2000).

An important product of rough set analysis are decision rules of an 'if... then...' nature. The method aims to determine which combinations of a classified set of attributes characterizing the objects are consistent with the occurrences of variation on the dependent variable or the decision attribute (Bruinsma et al., 2000).

For this analysis, we aim to describe the relationship between the decision attribute, the output multiplier, and ten condition attributes. The condition attributes are first considered as a group, and then divided into three sub-groups. The first of these sub-groups describes the characteristics of the information source, the 'meta-variables'; the second group describes the characteristics of the area, and the last group describes the characteristics of the tourism sector. Table 8.2 shows the relevant attributes per group. Attribute classes can be found in Appendix 8.I.

Table 8.2   Attributes per group

Information source	Area	Tourism
Year of data collection	Size of the *population*	*Year of data* collection
Type of *documentation*	Surface of the *area*	The number of *visitors*
	Geographic	Kind of *attraction* to tourists
	Kind of *attraction* to tourists	Total *expenditures* per year
	Population *density*	

8.6.2 The minimal set of reducts

First, we will examine the minimal set of reducts, together with their frequencies of appearance. The minimal subset of attributes, which ensures the same quality of classification as the set of all attributes, is called a 'reduct'. This minimal set of reducts contains no redundant information. It would be optimal if only one reduct were to occur, because the fewer possibilities for minimal sets, the higher the 'predictive power' of the information (Pawlak, 1991).

If an attribute appears in all the reducts it is called a 'core' attribute. This core attribute is the most meaningful attribute and the common part of all reducts.

According to van den Bergh et al. (1997), categorization of data is seen as the most problematic issue in taxonomic experiments. First of all, loss of information of continuous variables is involved, but another aspect of classification is that it can affect the outcomes. Different classifications can lead to different outcomes. Therefore, it is recommended that a sensitivity analysis be performed. Within this analysis we categorize the data in two similar ways according to what is called 'equal-frequency binning' (Witten and Frank, 2000). This method implies an even distribution of the attribute values over a predetermined number of bins; in this paper we distribute the sorted data over four bins and over two bins.

Table 8.3 describes the relative frequency of appearance of the condition attributes in the reducts. It shows that this appearance differs when using four, or two classes (bins). Because the number of reducts is smaller when using two classes, the predictive power is also higher. If we distinguish the four classes, no core attribute can be found. The condition attribute with the highest frequency of appearance when using four classes is 'density' and when using two classes 'geographic'. These two attributes are not clearly related, which indicates that classification is very important. On the other hand, we find that, in both cases, 'visitors', 'population' and 'expenditures' are relatively important.

If we derive the reducts for the groups of attributes, concerning the area, the tourism sector and the information source, we find that almost all the attributes are core attributes and therefore of equal importance.

Table 8.3   Minimal set of reducts of the output multiplier

Four equal groups		Two equal groups	
Attribute	*Frequency (%)*	*Attribute*	*Frequency (%)*
Density	95	Geographic	100
Visitors	58	Year of data	100
Area	53	Population	60
Population	47	Visitors	60
Expenditures	47	Expenditures	60
Geographic	42	Attraction	40
Type of documentation	42	Type of documentation	40
Attraction	32		
Year of data	32		

8.6.3 The decision rules

To obtain the decision rules we use the Rose program (Predki and Wilk, 1999) to calculate the basic minimal covering. We only use the rules with strength 2 or more. This means that the relation described in the rule appears at least twice in the data set, and in some cases it also appears eight times. Table 8.4 shows the decision rules for the output multipliers. For example, the first rule, when including all the attributes, means; IF the population attribute has value 2 and the area attribute has a value 3, THEN the income multiplier has a value 1. If we take a look at the classification of the data in Appendix 8.I, we can see that the decision rule can also be stated as: IF the size of the population is between 70,000 and 400,000 inhabitants, and the size of the area is between 5,000 km^2 and 25,000 km^2, THEN the income multiplier has a value between 1.10 and 1.32.

Table 8.4   Output multiplier rules when distinguishing four equal groups

		Population	Area	Visitors	Documentation	Year of data	Attraction	Geographic	Density	Multiplier
	1	2	3	–	–	–	–	–	–	1
Rules	2	–	–	–	2	–	4	3	3	1
related to	3	–	–	–	–	3	–	–	4	2
all the	4	–	–	–	3	–	–	–	3	3
indicators	5	4	–	–	3	–	–	–	–	4
	6	–	4	–	–	–	–	–	1	4

		Year of data	Documentation	Multiplier
Rules	1	–	–	1
related to	2	–	–	2
meta-	3	3	3	3
data	4	2	3	4
	5	1	2/3	4

		Population	Area	Attraction	Geographic	Density	Multiplier
	1	2	3	–	–	–	1
Rules	2	–	–	4	–	4	2
related to	3	4	–	–	–	2	4
the *area*	4	–	–	–	4	4	4
	5	–	4	–	–	1	4

		Visitors	Attraction	Expenditures	Year of data	Multiplier
Rules	1	2	–	–	4	1/2
related to	2	3	–	–	3	2/3
tourism	3	–	–	1	2	3/4
sector	4	–	4	–	2	4

As stated before, we first examine the relation between the decision attribute, the output multiplier, and all the condition attributes. Because only 24 cases are available and a lot of condition attributes are used, no obvious answers appear, but we can find some patterns. First of all, rules 1 and 5 show that a smaller population appears together with a low multiplier, and a larger population with a high multiplier. We can also find that type of documentation 2 (report) appears together with a low multiplier, and so do attraction 4 (mix) and geographic 3 (region). Furthermore, documentation 3 (conference paper) seems related to high multipliers.

If we take a look at the rules related to meta-data, some other patterns emerge. First of all, we find that older publications (1) relate to higher multipliers, and also that conference papers (3) often report higher multipliers. Unfortunately, there are no rules related to the lower multipliers. Recalling our three hypotheses, this again confirms the third hypothesis: the larger the time span for which the multiplier was derived, the higher the multiplier.

The rules related to the area state that a rather small population together with a medium-sized area (rule 1) may occur together with a low multiplier. On the other hand, a larger population with a relatively low density seems related to high multipliers. If we take a look at the density column, we see that different densities appear within one multiplier class. This is, of course, difficult to explain: there seems to be no clear relation. But, regarding rule 4, it appears that geographic 4, which means 'country' together with a high population density (higher than 150 inhabitants per km^2) is related to high multiplier values. This also confirms the hypothesis: the larger the economic base, the higher the multiplier.

Finally, we consider the tourism-related rules: they show us, in the first rule, that a relatively small number of visitors, reported in recent publications, leads to a low multiplier.

If we examine the attraction column, we find a contradiction with the rules related to the area and the rules related to all the attributes. The tourism-related rules find a statistical relation between attraction 4 (a mix of features) and high multipliers, while the other two groups of rules show that, if a mix of features attracts tourists, then low multipliers occur. The year of data gathering does confirm earlier statements that, if the data have been gathered earlier, the multipliers are higher.

So far, we have focussed on the rules derived while using four equal classes. As stated before, the classification can have large impacts on the outcomes of the analysis. Therefore, we will now examine the rules derived with the help of only two classes. Of course, the outcomes will be less diverse but they are also expected to be more clear.

First of all, these rules do not include the density or area attribute. The other set of rules (when using four classes) does include them, although the outcomes are not unambiguous.

When comparing the rules using four classes and using two classes, it appears that no (great) differences appear, except for the missing density attribute. This means that the outcomes can be referred to as 'robust'.

Concerning the attraction attribute, however, we now find that when visitors are attracted by a mix of characteristics and the data is relatively old, the multiplier is high. Unfortunately, no rules related to the tourism sector could be found.

Table 8.5   Output multiplier rules when distinguishing two equal classes

		Attraction	Geographic	Expenditure	Year of data	Documentation	Multiplier
Rules related to *all the indicators*	1	–	3	–	2	2	1
			1	2			1
	2	2	–	–	1		2
	3	–	4	–	–		2
	4	–	–	–	–	3	2

		Year of data	Documentation	Multiplier
Rules related to *meta-data*	1	–	–	1
	2	1	2	2
	3	–	3	2

		Population	Area	Attraction	Geographic	Multiplier
Rules related to *the area*	1	1	–	4	–	1
	2	–	–	–	4	2

		Visitors	Attraction	Expenditures	Multiplier
Rules related to *tourism sector*	1	–	–	–	–
	2	–	–	–	–
	3	–	–	–	–

8.7 Conclusions

In this chapter we performed a meta-analysis on tourism output multipliers. The basic research question we addressed in this chapter is: 'Which characteristics of the documentation source, the research area or the tourism sector affect the size of the output tourism multiplier?' The characteristics concerned have been divided into three sub-groups: the first group describes the characteristics of the documentation source (the 'meta-variables'); the second group describes the characteristics of the area; and the last group describes the characteristics of the tourism sector.

The first analysis considered average multiplier values according to relevant characteristics. First of all we found that, on average, the publication type 'report' and 'paper' have high multipliers. When focussing on the type of area, it appears that cities have low average multipliers and countries in particular show high values. Concerning the characteristics of the tourism sector, we find that high visitor numbers and expenditures are related to high multipliers. Furthermore, it appears that areas which are attractive because of nature and cultural values have high multipliers, while areas which are attractive because of a mix of factors have low values.

We then performed a linear regression analysis and for this we formulated three hypotheses:

1. The larger the economic base, the higher the multiplier.
2. The more visitors or expenditures of visitors, the higher the multiplier.
3. The longer the time span for which the multiplier was derived, the higher the multiplier.

We found that the country dummy and the population variable have a positive influence on the output multiplier, which agrees with the first hypothesis. The year of data has a negative impact, which confirms the third hypothesis. Unfortunately, we cannot say anything about the second hypothesis because the visitors variable showed no significant impacts.

The difference between the outcomes of linear regression and the rough set analysis is that the first of these methods indicates a potential causal relationship between the dependent and the independent variable, while rough set analysis refers to a statistical frequency at which a certain category of the decision attribute occurs in certain categories of the condition attributes. The derived rules are related to the three groups of attributes.

The rules related to meta-data show that older publications relate to higher multipliers and that conference papers often give higher multipliers. The rules related to the area state that a rather small population together with a medium-sized area often occur together with a low multiplier. It also appears that countries with a high population density are related to high multiplier values. Concerning the reason why people visit an area (attraction), we find relatively high multipliers when tourists are attracted to nature values but lower values when they are interested in a mix of characteristics.

Unfortunately, the rough set analysis also does not provide clear rules concerning the tourism sector. This means that the rough set rules do not substantially differ from the outcomes of the linear regression.

Trying to answer the basic research question in this paper, it appears first of all that the characteristics of the documentation source do have an effect on the size of the multiplier. According to all three analyses, we found that the multipliers reported in conference papers in particular have higher values than those reported in articles. We also found that recently derived multipliers often have lower values. This can be related to the increasing globalization taking place in many economies.

As can be found in many other publications, it appears that multipliers for countries are higher, and so are multipliers for areas with large populations. This indicates that, apart from the fact that higher multiplier values are expected for larger economies, national multipliers are higher than regional multipliers, probably due to the fact that boundaries decrease import leakage.

Furthermore, we could not find clear effects of the number of visitors or their expenditures on the output multipliers. This can be explained by the fact that tourism multipliers aim to derive the impact of visitors and their expenditures on the (local) economy, and that they therefore are not affected by them.

References

Armstrong, H. and J. Taylor (2000), *Regional Economics and Policy*, Blackwell, Oxford.

Baaijens, S.R. and P. Nijkamp (1996), 'Meta-analytic Methods for Comparative and Exploratory Policy Research: An Application to the Assessment of Regional Tourist Multipliers', *Journal of Policy Modelling*, 22:7, 821–58.

Baaijens, S.R., P. Nijkamp and K. van Montfoort (1997), 'Explanatory Meta-analysis of Tourist Income Multipliers: An Application of Comparative Research to Island Economics', *Tinbergen Institute Discussion Paper 97–017/3*, Rotterdam.

Bergh, J.C.J.M. van den, K.J. Button and P. Nijkamp (1997*)*, *Meta-analysis in Environmental Economics*, Kluwer Academic Publishers, Dordrecht.

Bruinsma, F., P. Nijkamp and R. Vreeker (2000), *Spatial Planning of Industrial Sites in Europe: A Benchmark Approach to Competitiveness Analysis About the Suitability of Sites for Economic Activities*, Research Memorandum 2000-7, Vrije Universiteit, Amsterdam.

Fletcher, J.E. (1989), 'Input-Output Analysis and Tourism Impact Studies', *Annals of Tourism Research*, 16: 514–29.

Frechtling, D.C. (1994), 'Assessing the Economic Impacts of Travel and Tourism – Introduction to Travel Economic Impact Estimation', in J.R. Brent Ritchie and C. R. Goeldner (eds), *Travel, Tourism and Hospitality Research*, John Wiley, New York.

Freeman, D. and E. Sultan (1999), 'The Economic Impact of Tourism in Israel: A Multi-regional Input-Output Analysis', in P. Rietveld and D. Shefer (eds), *Regional Development in an Age of Structural Economic Change*, Ashgate, Aldershot.

Miller, R. E. and P.D. Blair (1985), *Input-Output Analysis: Foundations and Extensions*, Prentice Hall, Englewood Cliffs.

Nijkamp, P. and G. Pepping (1998), 'Meta-analysis for Explaining the Variance in Public Transport Demand Elasticities in Europe', *Journal of Transportation and Statistics*, 1: 1–14.

Oltmer, K. (2003), *Agricultural Policy, Land Use and Environmental Effects – Studies in Quantitative Research Synthesis*, Dissertation, Tinbergen Institute, paper no. 318, Amsterdam.

Oosterhaven, J. and E.C. van de Knijff (1988), 'On the Economic Impacts of Recreation and Tourism: The Input-Output Approach', *Built Environment*, 13(2): 96–108.

Pawlak, Z. (1991), *Rough Sets: Theoretical Aspects of Reasoning About Data*, Kluwer Academic Publishers, Dordrecht.

Pindyck, R.S. and D.L. Rubinfeld (1991*)*, *Econometric Models and Economic Forecast*, McGraw-Hill, New York.

Predki, B. and S. Wilk (1999), 'Rough Set Based Data Exploration Using ROSE System', in Z.W. Ras and A. Skowron (eds), *Foundations of Intelligent Systems, Lecture Notes in Artificial Intelligence*, vol. 1609, Springer-Verlag, Berlin, pp. 172–80.

Stynes, D.J. (1997), *Economic Impacts of Tourism, A Handbook for Tourism Professionals*, Illinois Bureau of Tourism, Available on-line: <http://www.tourism.uiuc.edu/itn/etools/ eguides/econimpacts.pdf>.

Stynes, D.J. (1998), *Economic Impacts of Recreation and Tourism*, Available on-line: <http://www.msu.edu/course/prr/840/econimpact/multipliers.htm>.

Towner, J. (1995), 'What is Tourism's History?', *Tourism Management*, 16(5): 339–43.

United Nations (UN) Department for Economic and Social Information and Policy Analysis and World Tourism Organization (WTO) (1994), *Recommendations on Tourism Statistics*, United Nations, New York.

Witten, I.H. and E. Frank (2000), *Data Mining: Practical Machine Learning Tools and Techniques with Java Implementation*, Morgan Kaufman, San Francisco.

Appendix 8.I Attribute classes when using four equal classes and two equal classes

---

Four equal classes

Decision attribute	Classes
Output multiplier	
1	1.10–1.32
2	1.33–1.44
3	1.45–1.61
4	1.62–2.00

Condition attribute	Classes	Condition attribute	Classes
Population	(#)	Publication	
1	0 – 70,000	1	Article
2	70,000 – 400,000	2	Report
3	400,000 – 1,000,000	3	Conference Paper
4	> 1,000,000		
Area	(Km²)	Year of data	(Year)
1	0 – 600	1	1980–1993
2	600 – 5,000	2	1994–1996
3	5000 – 25,000	3	1997–1999
4	> 25,000	4	2000–2002
Visitors	(# /year)	Attraction	
1	0 – 800,000	1	Sun
2	800,000 – 1,300,000	2	Nature
3	1,300,000 – 10,000,000	3	Culture
4	> 10,000,000	4	Mix

Expenditures	(Mil, \$/year)	Geographic	
1	0 – 30	1	City
2	30 – 120	2	Island
3	120 – 1,000	3	Region
4	> 1,000	4	Country/State

Density	(Inh/km²)
1	0 – 10
2	10 – 40
3	40 – 150
4	> 150

Two equal classes

Decision attribute	Classes
Output multiplier	
1	1.10–1.44
2	1.45–2.00

Condition attribute	Classes	Condition attribute	Classes
Population	(#)	Publication	
1	0 – 300,000	1	Article
2	300,000 – 41,000,000	2	Report
		3	Conference Paper
Area	(Km²)	Year of data	(Year)
1	0 – 8,000	1	1980–1996
2	8,000 – 1,230,000	2	1997–2002
Visitors	(# /year)	Attraction	
1	0 – 1,400,000	1	Sun
2	1,400,000 – 92,000,000	2	Nature
		3	Culture
		4	Mix
Expenditures	(Mil, \$/year)	Geographic	
1	0 – 200	1	City
2	200 – 10,000	2	Island
		3	Region
		4	Country/State
Density	(Inh/km²)		
1	0 – 40		
2	40 – 4,000		

Chapter 9

Competition among Tourist Destinations: An Application of Data Envelopment Analysis to Italian Provinces

M. Francesca Cracolici and Peter Nijkamp

9.1 Setting the scene

The leisure industry has become a prominent economic sector in the Western world. More discretionary income and more free time have created the foundation for a new lifestyle in our society where recreation and tourism make up major elements of daily behaviour. In many regions and countries, tourism is regarded as one of the major growth vehicles. Despite crowding effects and many other negative externalities, it seems plausible to assume that tourism (and recreation) will remain an important growth industry. Along with the increasing economic importance of the tourist sector, we also witness increasing competition on the tourist market. Such competition has to seek a balance between short-term revenues at the cost of long-term sustainable development and long-term balanced growth strategies which seek to reconcile local interests with tourism objectives. In practice, we observe that different tourist destinations try to exploit their indigenous potential. This requires a well-tuned marketing strategy in order to get the right tourist with the right goals at the right place (see Coccossis and Nijkamp, 1995). In this context, Buhalis and Fletcher (1992, p. 10) quote Goodall who claimed that: 'the demand of increasing numbers of tourists is satisfied in a manner which continues to attract them whilst meeting the needs of the host population with improved standards of living, yet safeguarding the destination environment and cultural heritage'.

Leisure time activities have indeed assumed a prominent place in the activity patterns of many people. At the same time, we observe that the time spent on discretionary activities is very scarce, so that time competition is a real issue. This will have a far-reaching impact on tourism, as more people wish to go more often on vacation, though usually for a shorter period.

Tourism has also become a global activity. The new trend in world tourism towards non-traditional and remote destinations is an expression of the passage from mass tourism to a *new age of tourism*, and points to a change in the attitudes and needs of tourists (Fayos-Solá, 1996; Poon, 1993). Distant or previously unknown destinations have become places to explore, since they are potentially able to supply what the tourist expects, viz. a total leisure experience.

The tourist destination tends to be seen not as a set of distinct natural, cultural, artistic and environmental resources, but as an overall product, a complex and integrated package offered by a territory able to supply a holiday which meets the varied needs of the tourist. All this leads to strong competition between traditional destinations seeking to maintain and expand their market share and new destinations trying to earn a significant market share. This competition is centred not on the single aspects of the tourist product (environmental resources, transportation, tourism services, hospitality, etc.), but in particular on the destination as a unifying and central factor of the tourist system (Buhalis, 2000; Crouch and Ritchie, 2000). As a consequence, destinations face the challenge to manage and organize their resources in order to supply a holiday experience that must be equal to, or better than, the alternative destination experiences on the market.

From this comes awareness, by nations, regions, provinces and cities, of the role of the territory as a tourist destination: namely, an integrated amalgam of tourist resources (natural and cultural resources, transportation, tourism services, hospitality, etc.) able to supply a satisfactory product that tourists appreciate (Buhalis, 2000).

According to destination marketing management, the above considerations imply that destination management organizations (DMOs) need to strive to develop tourism policies based on strategies and operating actions that give them an advantage over their competitors. As a consequence, it is important to be able to measure the performance of each area against its 'key competitors' in order to identify proper strategic actions needed to maintain or strengthen its position as a market leader. In the light of these factors, aspects such as destination tourist performance, destination competitiveness and its measurement have attracted increasing interest in the tourism literature in recent years (Alavi and Yasin, 2000; Crouch and Ritchie, 1999, 2000; Kozak and Rimmington, 1999; Kozak, 2002).

This chapter focusses on the concept of tourist destination competitiveness with special attention to a measurement system using Data Envelopment Analysis (DEA) (Charnes et al., 1978) that is applied to the 103 Italian provinces. In particular, we will provide a measurement of competitiveness in terms of efficiency; in other words, using the metaphor of the territory as a firm, we hypothesize that the resources (material and human) in the territory constitute the input of a *virtual production process*, the output of which is tourist flows. In the light of this, destination tourist performance can be evaluated by the capability of a territory to transform its stock resources into maximum production. That is to say, productive efficiency of a territory to produce tourist flows can be viewed as a proxy for destination competitiveness. This new idea will be further investigated in the upcoming chapter.

9.2 Tourist destination competitiveness: Analytical framework

Competitiveness plays a key role in our study. But what do we mean by 'competitiveness'? What is 'destination competitiveness'? And, what are the strategic factors determining destination competitiveness?

The concept of competitiveness can seem easily understandable – it is the expression of the qualitative and quantitative superiority of a unit (a firm, a territory) over the real and potential competitor set. However, the complexity of the concept is made evident when we seek to define and measure it, as is shown by several literature sources. For example, Porter (1990) argues that its complexity comes from the wide variety of perspectives on competitiveness, which makes it difficult to give an exhaustive and universal definition. Scott and Lodge (1985) connect this complexity to the multidimensional and relative nature of the concept of competitiveness. The versatile nature of competitiveness concerns its essential qualities, while the relative aspect deals with the concept of superiority; but superior in comparison to what and to whom?

These considerations have led to a proliferation of definitions of competitiveness over the years. Scott and Lodge (1985) define competitiveness between countries as: 'a country's ability to create, produce, distribute, and/or service products in international trade while earning rising returns on its resources'. They identify this ability more as a strategic factor than as something related to natural endowments.

In 1992, Newall viewed competitiveness as an essential and key factor in achieving national prosperity. He argued that competitiveness:

...is about producing more and better quality goods and services that are marketed successfully to consumers at home and abroad. It leads to well paying jobs and to the generation of resources required to provide an adequate infrastructure of public services and support for the disadvantaged. In other words, competitiveness speaks directly to the issue of whether a nation's economy can provide a high and rising standard of living for our children and grandchildren (p. 94).

In 1994, the Organization for Economic Cooperation and Development (OECD) defined competitiveness as: 'the degree to which a country can, under free and fair market conditions, produce goods and services which meet the test of international markets, while simultaneously maintaining and expanding the real income of its people over the long term' (p. 18). Putting it another way, the OECD speaks of competitiveness as: 'the ability of a country or company to, proportionally, generate more wealth than its competitors in world markets' (OECD, 1994, p. 18).

The above considerations and definitions underline that it is not simple to give an overall definition for such a complex concept as competitiveness. In particular, the previous definitions show different perspectives of competitiveness and the different areas where competition exists: at company level, national industry level, and national economy level (Crouch and Ritchie, 1999). To which of these levels does destination competitiveness belong? On what factors is competition between destinations based? Why do destinations with fewer resources perform better than those endowed with better resources?

Crouch and Ritchie (1999) argue that tourist destination competitiveness fits into the national industry competition level. These authors provide a detailed framework in which the different perspectives on competitiveness are coherently organized. In their opinion, in particular, the framework of Porter's diamond model 'suggests the fundamental structure of competition among national tourism industries' (Crouch and Ritchie, 1999, p. 140). Following Porter's (1990) diamond model, competition between destinations at different levels (country, region, province, etc.) can be considered in terms of the determinants of territorial tourist advantage.

The national diamond model identifies six elements on which competition among national industries – or destinations, in the case of tourism – is based. These elements are: factor conditions; demand conditions; related and supporting industries; firm strategy, structure, and rivalry; chance events; and government.

Factor conditions are the set of natural and man-made resources. They are the strategic elements of destination competition, because they are the main motivational factors of a holiday.

Demand conditions are fundamental for a long-term position in the market, especially a national demand market, because a good domestic position works towards improvement and innovation processes which develop and strengthen the international market position in the long-run.

The intersectorial nature of tourism, where production is the integrated result of different operators, assigns an important role to the *related and supporting industries* because their good performance can give a competitive edge.

Added to this, in Porter's opinion, a competitive entrepreneurial environment (*firm strategy, structure, and rivalry*) avoids stagnation and stimulates improvement.

Finally, the 'diamond model' considers two uncontrollable variables that indirectly influence national industries, in this case the tourism industry; they are *chance events* – that can be, depending on the circumstances, an opportunity or a threat – and the *government*.

To sum up, there are a large number of different elements on which competition among tourist destinations is based. Therefore, achieving a good performance and position in the market depends on the capability of a destination to manage and organize its resources according to a systematic logic. In particular, because of the heterogeneity of resources and the plurality of operators involved in the tourism sector, destination management organizations (DMOs) should aim to develop a systematic approach to tourism policy. Therefore, in tourism, though the factor conditions are strategic elements and are necessary in order to compete, the questions arise: Are they enough on their own? Do only endowment resources represent competitive strength? Or, is it the capability of a destination to organize and manage its resources which enables it to achieve a market leader position, a competitive advantage?

Obviously, physical, natural and cultural resources are important factors of a tourist destination's competitiveness, but differing market positions and market-shares among similar tourist destinations lead to the conclusion that the determinants of competition are complex. It is *how* the territorial endowments are managed that allows a destination to create a *competitive advantage* (Porter, 1990).

Crouch and Ritchie argue that the opportunity for a destination to create a competitive advantage depends on five elements: *audit* or *inventory of resources, maintenance of resources; growth and development of the stock of resources; efficiency or resource management;* and *effectiveness of resources deployment* (Crouch and Ritchie, 1999, p. 144).

In the next section, we focus on the capability of a destination to use its tourist resources efficiently, using Data Envelopment Analysis (DEA) (Charnes et al., 1978) to estimate how well Italian provinces utilize their tourist resources. Efficiency scores, as a proxy for competitiveness, are used to rank the Italian provinces in two clusters: 'efficient' and 'inefficient' provinces. In relation to this, our purpose is to measure the technical efficiency of Italian destinations. We also aim to examine whether there are efficiency differences among Italian provinces in producing tourist flows.

9.3 Data envelopment analysis

If the territory is analyzed as if it were a company, then we can hypothesize that a tourist area should be able to manage its input efficiently: in other words, the territory's physical and human resources constitute the input of a virtual tourist *'production process'*, and the output is arrivals, bed-nights, value added, employment, customer satisfaction, etc. As a consequence, destination performance can be evaluated through a measurement of efficiency. A territory can have its performance evaluated by measuring its efficiency thus allowing destination managers to check and identify any process dysfunctions in order to define strategic and operational actions.

Following this hypothesis, where the production function is not known, the efficiency of a destination can be evaluated against that of its main competitors. A destination will be efficient if all its input-output (respectively physical and human resources input and tourist output) combinations are better than those of its identified 'key competitors'. In view of all these considerations, the following 'guest-production function' for tourism is proposed:

Tourist output = f (material capital, cultural heritage, human capital, labour)   (1)

Tourist output is evaluated by a non-financial measure: international and national bed-nights. According to the destination concept and the empirical findings, and the availability of data, the following proxies for material capital, cultural heritage, human capital, and labour were chosen: number of beds in the hotels and in complementary accommodation divided by population; the regional State-owned artistic patrimony (number of museums, monuments and archaeological areas) divided by population; tourist school graduates divided by working age population; and the labour units (ULA) of the tourism sector divided by the total regional ULA.

Given the production process function (1), a measurement of technical efficiency expresses the capability of different destinations to transform input (tourist resources) into output (tourist flows). For this purpose, because the functional form of the 'guest-production function' is not known, data envelopment analysis (DEA) (Charnes et al., 1978) is adopted using the above multiple inputs and outputs.

DEA is a non-parametric linear programming method of measuring efficiency to determine a production frontier. The efficiency of each tourist destination is evaluated against this frontier. In other words, the efficiency of a destination is evaluted in comparison with the performance of other destinations.

DEA is based on Farrell's (1957) work, further elaborated by Charnes et al.'s (1978) CCR Model and Banker et al.'s (1984) BCC Model. It has been widely used in empirical efficiency analysis because it does not require an assumption about functional form, and it can be used in cases where the units (Decision-Making Units, DMUs) use multiple inputs to produce multiple outputs.

Generally, it is applied in public sector agencies (e.g. schools, hospitals, airports, courts, etc.) and private sector agencies (banks, hotels, etc.). Our analysis aims at assessing the tourist efficiency of territorial areas. In recent years, several regional applications of DEA have emerged. Macmillan (1986) used DEA to assess the efficiency of cities in China. Charnes et al. (1989) employed DEA to evaluate the industrial performance of 28 Chinese cities in 1983 and 1984. More recently, Susiluoto and Loikaanen (2001) used DEA to assess the economic efficiency of Finnish regions in 1988 and 1999. Martić and Savić (2001) used DEA to evaluate how well regions in Serbia utilize their resources in regard to socio-economic development. Cuffaro and Vassallo (2002), using data from the World Bank, employed DEA to evaluate the efficiency of 83 countries in transforming their resources into human development in 1985 and 1995.

We use data for two outputs and five inputs, in order to estimate how well provinces in Italy utilize their tourist resources. For this purpose, we adopt an output-oriented constant returns-to-scale DEA model (CCR Model) for each province in 1998 and 2001. We deploy such a model because our aim is to explore how well Italian provinces utilize their tourist resources. In other words, given a stock of tourist resources, the aim is to maximize tourist flows. It is plausible that constant returns to scale prevail in many operations of the regions.

The efficiency measure proposed by Charnes et al. (1978) maximizes efficiency in terms of the ratio of total weighted output to total weighted input, subject to the condition that, for every destination, this efficiency measure is smaller than or equal to 1. Given n destinations with m inputs and s outputs, the measure of efficiency of a destination k can be specified as:

$$\underset{u,v}{\text{Max}} \ \frac{\sum_{r=1}^{s} u_r y_{rk}}{\sum_{i=1}^{m} v_i x_{ik}}$$

$$\text{s.t.} \ \frac{\sum_{r=1}^{s} u_r y_{rj}}{\sum_{i=1}^{m} v_i x_{ij}} \leq 1; \ \text{for } j=1,...,n, \tag{2}$$

$$v_i, u_r \geq 0,$$

where x_{ij} is the amount of input i to destination j; y_{rj} the amount of output r from destination j; u_r the weight given to output r; and v_i the weight given to input i.

The maximization problem in (2) can have an infinite number of solutions. Charnes et al. (1978) show that the above fractional programming problem has the following linear programming equivalent, which avoids this problem:

$$\underset{u,v}{\text{Max}} \ \sum_{r=1}^{s} u_r y_{rk}$$

$$\text{s.t.} \ \sum_{i=1}^{m} v_i x_{ij} - \sum_{r=1}^{s} u_r y_{rj} \geq 0; \ \text{for } j=1,...,n,$$

$$\sum_{i=1}^{m} v_i x_{ik} = 1; \tag{3}$$

$$u_r \geq 0; \ \text{for } r=1,...,s,$$

$$v_i \geq 0; \ \text{for } i=1,...,m.$$

The dual problem of this linear programming can be written as follows:

$$\underset{\theta,\lambda}{\text{Min}} \ \theta_k$$

$$\text{s.t.} \ \sum_{j=1}^{n} \lambda_j y_{rj} \geq y_{rk}; \ \text{for } r=1,...,s, \tag{4}$$

$$\theta_k x_{ik} - \sum_{j=1}^{n} \lambda_j x_{ij} \geq 0; \ \text{for } i=1,...,m,$$

$$\lambda_j \geq 0; \ \text{for } j=1,...,n.$$

The destination, j, is efficient, if $\theta^* = 1$, where an asterisk to a variable denotes its optimal solution. If this condition is not satisfied, the destination j is inefficient ($\theta^* > 1$).

Andersen and Petersen's (1993) A&P Model proposed a modified model of (3) to increase the discrimination power for every efficient destination by adding a constraint, $j \neq k$.

The basic idea of the model is to compare the unit under evaluation with a linear combination of all other units in the sample, excluding the DMU itself. The efficiency score obtained reflects the radial distance between the production frontier, evaluated without (efficient) unit k, and the unit itself, i.e. for $J = \{j = 1,...n, j \neq k\}$.

Given n destinations with m inputs and s outputs, the measure of super-efficiency of a destination, k, can be obtained by solving the following model:

$$\underset{u,v}{\text{Max}} \sum_{r=1}^{s} u_r y_{rk}$$

$$\text{s.t.} \sum_{i=1}^{m} v_i x_{ij} - \sum_{r=1}^{s} u_r y_{rj} \geq 0; \quad \text{for } j=1,...,n, j \neq k,$$

$$\sum_{i=1}^{m} v_i x_{ik} = 1,$$

$$u_r \geq 0; \quad \text{for } r=1,...,s,$$

$$v_i \geq 0; \quad \text{for } i=1,...,m.$$

$$(5)$$

The dual formulation of this model is:

$$\underset{\theta,\lambda}{\text{Min}} \ \theta_k$$

$$\text{s.t.} \sum_{\substack{j=1 \\ j \neq k}}^{n} \lambda_j y_{rj} \geq y_{rk}; \quad \text{for } r=1,...,s,$$

$$\theta_k x_{ik} - \sum_{\substack{j=1 \\ j \neq k}}^{n} \lambda_j x_{ij} \geq 0; \quad \text{for } i=1,...,m,$$

$$\lambda_j \geq 0; \quad \text{for } j=1,...,n, j \neq k.$$

$$(6)$$

Having now specified the formal model for evaluating the performance of actors (i.e., provinces), we will now turn to the actual application in Italy.

9.4 Empirical findings

In this section, the DEA results, obtained by the EMS software (Scheel, 2000), will be presented and discussed. Data on outputs have been obtained from ISTAT (National Statistics Institute) (1998 and 2001) Tourist Statistics, while the data on inputs has been obtained from different sources: number of beds in the hotels and complementary accommodation from ISTAT Tourist Statistics (1998 and 2001);

provincial State-owned artistic heritage (number of museums, monuments and archaeological areas) from the Ministry of Cultural Heritage[1]; tourist school graduates from the Ministry of Education; and labour units (ULA) of the tourism sector from ISTAT provincial accounts (1998 and 2000[2]).

The DEA results show slight differences between Italian tourist destinations in the two years considered. In 1998 there were 32 efficient provinces (Table 9.1) and the average technical efficiency appears to be high (0.78), while by 2001 the number of efficient destinations had decreased. Then, only 29 provinces achieved an efficiency coefficient of 1 (Table 9.2) and the average technical efficiency is almost equal to 1998 (0.76).

In 2001 among the provinces with a decreased performance, compared with 1998, were: Aosta, Cremona, Mantua, Pordenone, Trieste, Parma, Massa-Carrara and Ancona. Among these destinations, only the provinces of Cremona and Mantua show an efficiency score lower than the average technical efficiency. Meanwhile, among the destinations with a better performance – with the exception of Trento (a typical mountain destination) – are business areas or coastal and cultural destinations such as Milan, Macerata, Teramo, and Pescara.

The top positions – for both years – among coastal destinations were held by Rimini, Imperia and Salerno, while in 2001, in comparison with 1998, a great deal of competitiveness was lost by coastal destinations such as Crotone, Vibo Valentia, Reggio Calabria and Messina. The best mountain and lake destinations – for both the years – were Verbano-Cusio-Ossola, Bolzano-Bozen and Belluno, with Trento being added in 2001. The other efficient destinations are typical business areas like Padua, Savona, Prato, Livorno, etc.

Among cultural and artistic destinations that maintain their efficiency position over the two years are Verona, Venice, Florence and Pisa, while Rome (historical and cultural tourist destination), though inefficient, shows an increased performance with respect to 1998.

The inefficiency of traditional tourist destinations (coastal and cultural destinations, i.e., Rome, Naples, Messina, Sassari, Nuoro, etc.) can be interpreted by an overendowment of inputs in relation to their production of tourist flows and their 'key competitors', or it could be an expression of the mature phase of popularity and saturation and/or the beginning of the phase of fading popularity of the life cycle of the Italian tourist product (cultural and coastal) (Prosser, 1994).

---

[1] Because the statistics from the Ministry of Cultural Heritage do not supply the data of regions and provinces with special statute status (Sicily, Aosta, Trento and Bolzano), for this observation we have used as a proxy for the cultural heritage the Region and Province-owned artistic heritage (museums, monuments and archaeological areas) (1998 and 2001) supplied by the Regional and Provincial Bureaus of Cultural Heritage.

[2] Data for the year 2001 are not available, so we have used the nearest available (2000). In addition, ULA also includes: commerce, repairs, hotels, restaurants, transport and comunication. If the indirect impact of tourism on commerce and repairs is considered, any error with this variable can be discounted.

In the light of these considerations, the inferior performance of the traditional destinations means they must improve the management of their endowments. Tourist resources represent a stock that does not decrease, so better efficiency is needed to increase tourist bed-nights. The stereotypical 'cultural and coastal' destination is no longer able to attract substantial tourist flows. To become more attractive, such a destination has to show that the Italian product is different from the other tourist products on the market.

We next tried to explore the cluster of efficent destinations, using the A&P model for both years (Andersen and Petersen, 1993). There were two outliers, Avellino and Lodi, with low super-efficiency scores.[3] The reason for this can be found in the composition of the inputs and outputs of these provinces.

An analysis of inputs and outputs shows, on the one hand, that the value of inputs for these provinces is lower than that for the other provinces and, on the other hand, the proportion of inputs and outputs in comparison with other provinces is high.

This indicates that there is a lack of homogeneity in the set of the provinces under consideration, probably due to different tourist functions among the Italian destinations, or to the absence for some destinations of a tourist function. The latter may be the case for Lodi and Avellino.

Because of that, we assessed all the destinations' efficiency without the presence of the provinces of Lodi (18) and Avellino (77), but did not observe any significant change in the results. In 1998, the efficient destinations cluster, based on the CCR model, maintained its composition, with Milan being added because of the absence of Lodi and Avellino. It is known that, in the case of constant returns to scale, the values for weight λ_j are influenced by the scale size of the efficient units. Therefore, when provinces 18 and 77 are excluded from the evaluation, the change of the efficiency of Milan depends on the share of weight λ_{18} and λ_{77} in the sum of all λ_j.

The efficiency coefficient of Milan was 0.96 when the provinces of Lodi (18) and Avellino (77) were included in the evaluation. In this case, its peer group included the provinces of Lodi ($\lambda_{18} = 0,60$), Padua ($\lambda_{28} = 0,32$) and Avellino ($\lambda_{77} = 0.09$), and from this it follows that especially the share of weight of λ_{18} could not be discounted.

With regard to the A&P model assessed without provinces 18 and 77, the super-efficiency score changed only for the provinces of Cremona and Mantua whose score appeared to decrease (Table 9.3).

---

[3]   The efficiency coefficients for the A&P Model are reported in Tables 9.3 and 9.4; they concern the provinces set without Lodi and Avellino. Because of space limitations, we did not report the efficiency coefficients for the A&P Model with Lodi and Avellino.

Table 9.1 DEA efficiency results[4] (CCR model), 1998

	Province	Score	Benchmarks		Province	Score	Benchmarks
1	Torino	0.5345	34, 51, 78	53	Pisa	1	5
2	Vercelli	0.5505	35, 47	54	Arezzo	0.6251	34, 50, 51, 78
3	Biella	0.7119	35, 47	55	Siena	0.8624	21, 27, 28, 34, 46, 50
4	Verbano-Cusio-Ossola	1	1	56	Grosseto	1	10
5	Novara	0.8484	32, 35, 47	57	Perugia	0.6125	34, 50, 51, 78
6	Cuneo	0.3587	34, 51, 78	58	Terni	0.6392	34, 51, 60, 95
7	Asti	0.4193	34, 50, 51, 78	59	Pesaro e Urbino	0.7194	35, 56, 60
8	Alessandria	0.3557	20, 34, 39, 51	60	Ancona	1	15
9	Aosta	1	1	61	Macerata	0.7283	29, 56, 66
10	Varese	0.7199	26, 28, 34, 51, 78	62	Ascoli Piceno	1	1
11	Como	0.6496	4, 30, 32, 35	63	Viterbo	0.5885	56, 60, 66
12	Lecco	0.6565	26, 32, 35	64	Rieti	0.3675	18, 19, 35, 46
13	Sondrio	0.7262	20, 34, 39, 46, 51	65	Roma	0.8758	28, 50, 51
14	Milano	0.9599	18, 28, 77	66	Latina	1	4
15	Bergamo	0.6440	19, 26, 35, 46	67	Frosinone	0.6587	20, 28, 46, 51
16	Brescia	0.8807	23, 27, 32, 78	68	L'Aquila	0.6047	34, 35, 46, 95
17	Pavia	0.8237	18, 19, 20, 34, 46	69	Teramo	0.9219	33, 35, 56
18	Lodi	1	5	70	Pescara	0.6888	20, 34, 39, 46
19	Cremona	1	8	71	Chieti	0.4437	34, 51, 60, 78
20	Mantova	1	14	72	Isernia	1	0
21	Bolzano-Bozen	1	5	73	Campobasso	0.5696	34, 51, 60, 95
22	Trento	0.9673	9, 21, 33, 35, 46	74	Caserta	0.5149	34, 50, 51, 78
23	Verona	1	1	75	Benevento	0.6741	18, 19, 20, 95
24	Vicenza	0.8134	26, 30, 35, 46	76	Napoli	0.8780	26, 28, 50, 51, 78
25	Belluno	1	1	77	Avellino	1	1
26	Treviso	1	5	78	Salerno	1	16
27	Venezia	1	6	79	Foggia	0.7778	29, 56, 62
28	Padova	1	13	80	Bari	0.4738	19, 20, 34, 51
29	Rovigo	1	6	81	Taranto	0.6835	34, 51, 78
30	Pordenone	1	2	82	Brindisi	0.6149	34, 51, 78
31	Udine	0.7706	27, 29, 32, 53, 56	83	Lecce	0.8589	35, 47
32	Gorizia	1	5	84	Potenza	0.6599	20, 34, 39, 51
33	Trieste	1	3	85	Matera	0.6158	53, 60, 66, 78
34	Imperia	1	38	86	Cosenza	0.3644	35, 53, 56, 60
35	Savona	1	22	87	Crotone	0.4429	35, 56, 60
36	Genova	0.8064	19, 34, 51	88	Catanzaro	0.4378	29, 35, 60
37	La Spezia	0.7235	34, 50, 53, 78	89	Vibo Valentia	0.5255	25, 35, 56
38	Piacenza	0.4779	20, 28, 34, 51, 95	90	Reggio di Calabria	0.3187	29, 66
39	Parma	1	4	91	Trapani	0.6157	34, 51, 60, 95
40	Reggio nell'Emilia	0.9238	20, 34, 46, 95	92	Palermo	0.7868	20, 28, 34, 51, 95
41	Modena	0.6826	19, 20, 34, 46	93	Messina	0.9181	21, 28, 34, 46, 95
42	Bologna	0.9946	18, 19, 20, 28, 34, 46	94	Agrigento	0.6645	34, 51, 60
43	Ferrara	0.9303	27, 29, 50, 53, 78	95	Caltanissetta	1	14
44	Ravenna	0.8239	21, 27, 35	96	Enna	0.6634	20, 28, 34, 46, 95
45	Forli'-Cesena	0.7564	34, 35, 46	97	Catania	0.7479	34, 51, 60, 95
46	Rimini	1	19	98	Ragusa	0.5144	34, 51, 60, 78, 95
47	Massa-Carrara	1	4	99	Siracusa	0.7561	34, 51, 60, 95
48	Lucca	0.6253	27, 28, 34, 46, 50	100	Sassari	0.6059	34, 35, 60
49	Pistoia	0.8942	21, 28, 46, 50	101	Nuoro	0.6698	35
50	Firenze	1	11	102	Oristano	0.6055	33, 56
51	Prato	1	27	103	Cagliari	0.7205	34, 35, 46, 95
52	Livorno	1	0				

---

[4]   For efficient units, the benchmark column shows for efficient units the frequency with which an efficient unit appears in different groups, while, for inefficient units it shows the peer group.

Table 9.2   DEA efficiency results[5] (CCR model), 2001

	Province	Score	Benchmark			Province	Score	Benchmark	
1	Torino	0.5047	18, 34, 50, 78		53	Pisa	1		2
2	Vercelli	0.4761	18, 32, 35		54	Arezzo	0.5989	18, 34, 50, 78	
3	Biella	0.5958	18, 32		55	Siena	0.9601	21, 22, 27, 46	
4	Verbano-Cusio-Ossola	1		3	56	Grosseto	1		0
5	Novara	0.6888	18, 32, 35,		57	Perugia	0.8252	27, 34, 50, 78	
6	Cuneo	0.3426	18, 34, 35, 78		58	Terni	0.8500	18, 35, 66, 78	
7	Asti	0.3072	18, 50, 51, 78		59	Pesaro e Urbino	0.7318	25, 35, 52, 62	
8	Alessandria	0.3910	18, 34, 35, 46		60	Ancona	0.8149	35, 62, 78,	
9	Aosta	0.9013	22, 25, 46		61	Macerata	1		0
10	Varese	0.5799	18, 23, 29, 50		62	Ascoli Piceno	1		7
11	Como	0.9394	4, 46		63	Viterbo	0.2369	29, 52, 66, 78	
12	Lecco	0.6754	4, 18, 26, 32, 35		64	Rieti	0.4854	18, 35, 46	
13	Sondrio	0.7037	18, 28, 34, 46		65	Roma	0.9365	18, 28, 34, 50	
14	Milano	1		0	66	Latina	1		6
15	Bergamo	0.5929	18, 26, 35, 46		67	Frosinone	0.6237	18, 28, 34, 46	
16	Brescia	0.7792	23, 28, 34, 50, 78		68	L'Aquila	0.7407	18, 35, 46	
17	Pavia	0.6006	18, 28, 34, 46		69	Teramo	1		1
18	Lodi	1		46	70	Pescara	1		0
19	Cremona	0.6007	18, 50, 51		71	Chieti	0.5800	18, 35, 66, 78	
20	Mantova	0.6839	18, 28, 34, 46		72	Isernia	1		2
21	Bolzano-Bozen	1		3	73	Campobasso	0.6414	18, 35, 78	
22	Trento	1		3	74	Caserta	0.5736	29, 50, 51, 78	
23	Verona	1		3	75	Benevento	0.4276	18, 66	
24	Vicenza	0.6974	18, 32, 35		76	Napoli	0.8985	18, 28, 50	
25	Belluno	1		5	77	Avellino	1		0
26	Treviso	1		3	78	Salerno	1		28
27	Venezia	1		8	79	Foggia	0.6390	25, 29, 62, 66	
28	Padova	1		13	80	Bari	0.4557	18, 35, 78	
29	Rovigo	1		6	81	Taranto	0.5871	18, 35, 78	
30	Pordenone	0.9108	4, 26, 46		82	Brindisi	0.5272	18, 35, 78	
31	Udine	0.8214	27, 29, 32		83	Lecce	0.9833	18, 32	
32	Gorizia	1		10	84	Potenza	0.4727	18, 35, 46	
33	Trieste	0.8127	27, 32, 34, 50		85	Matera	0.6596	35, 62, 78	
34	Imperia	1		27	86	Cosenza	0.4531	35, 62, 78	
35	Savona	1		32	87	Crotone	0.2895	35, 62, 78	
36	Genova	0.7885	18, 28, 34, 50		88	Catanzaro	0.3907	18, 35, 62, 78	
37	La Spezia	0.7003	34, 50, 53, 78		89	Vibo Valentia	0.3421	25, 27, 35, 46	
38	Piacenza	0.4861	18, 34, 50, 78		90	Reggio di Calabria	0.2613	18, 66	
39	Parma	0.9686	18, 35, 46		91	Trapani	0.6390	18, 35, 78	
40	Reggio nell'Emilia	0.8333	18, 34, 35, 46		92	Palermo	0.7746	18, 28, 34, 50	
41	Modena	0.5897	18, 34, 35, 46		93	Messina	0.8330	28, 34, 46, 50	
42	Bologna	0.9206	18, 28, 34, 46		94	Agrigento	0.7059	18, 34, 50, 78	
43	Ferrara	0.8919	29, 50, 53, 78		95	Caltanissetta	1		1
44	Ravenna	0.8701	25, 27, 35, 46		96	Enna	0.4859	18, 50, 51, 78	
45	Forlì'-Cesena	0.8721	21, 22, 35, 46, 72		97	Catania	0.7381	18, 34, 50, 78	
46	Rimini	1		24	98	Ragusa	0.5609	18, 34, 50, 78	
47	Massa-Carrara	0.9645	32		99	Siracusa	0.7556	18, 34, 50, 78	
48	Lucca	0.6538	23, 28, 34, 46, 50		100	Sassari	0.6583	27, 34, 35, 50	
49	Pistoia	0.9251	21, 28, 46		101	Nuoro	0.6785	32, 35	
50	Firenze	1		24	102	Oristano	0.8773	27, 69, 72, 95	
51	Prato	1		4	103	Cagliari	0.7407	18, 34, 35, 46	
52	Livorno	1		2					

5   See Footnote 3.

Table 9.3   DEA efficiency results (A&P model), 1998

	Province	Score			Province	Score
33	Trieste	0.0753		45	Forli'-Cesena	1.3221
46	Rimini	0.3006		99	Siracusa	1.3226
4	Verbano-Cusio-Ossola	0.3556		97	Catania	1.3370
28	Padova	0.3938		61	Macerata	1.3731
72	Isernia	0.4470		13	Sondrio	1.3770
21	Bolzano-Bozen	0.4992		37	La Spezia	1.3821
95	Caltanissetta	0.5464		103	Cagliari	1.3879
35	Savona	0.5963		10	Varese	1.3892
20	Mantova	0.6378		59	Pesaro e Urbino	1.3900
27	Venezia	0.6495		3	Biella	1.4046
32	Gorizia	0.6600		70	Pescara	1.4519
29	Rovigo	0.6873		75	Benevento	1.4578
26	Treviso	0.7137		81	Taranto	1.4630
19	Cremona	0.7172		41	Modena	1.4649
50	Firenze	0.7227		101	Nuoro	1.4930
14	Milano	0.7328		94	Agrigento	1.5048
30	Pordenone	0.7426		96	Enna	1.5073
51	Prato	0.7726		84	Potenza	1.5154
23	Verona	0.7913		67	Frosinone	1.5182
56	Grosseto	0.8399		12	Lecco	1.5232
53	Pisa	0.8648		11	Como	1.5395
34	Imperia	0.8901		15	Bergamo	1.5528
78	Salerno	0.9088		58	Terni	1.5644
60	Ancona	0.9202		48	Lucca	1.5992
9	Aosta	0.9207		54	Arezzo	1.5998
66	Latina	0.9593		85	Matera	1.6239
62	Ascoli Piceno	0.9677		91	Trapani	1.6242
39	Parma	0.9704		82	Brindisi	1.6264
52	Livorno	0.9707		57	Perugia	1.6327
47	Massa-Carrara	0.9766		100	Sassari	1.6504
25	Belluno	0.9803		102	Oristano	1.6515
42	Bologna	1.0003		68	L'Aquila	1.6538
22	Trento	1.0339		63	Viterbo	1.6993
43	Ferrara	1.0750		73	Campobasso	1.7557
40	Reggio nell'Emilia	1.0825		2	Vercelli	1.8164
69	Teramo	1.0847		1	Torino	1.8710
93	Messina	1.0892		89	Vibo Valentia	1.9031
49	Pistoia	1.1183		74	Caserta	1.9422
16	Brescia	1.1354		98	Ragusa	1.9441
76	Napoli	1.1389		38	Piacenza	2.0924
65	Roma	1.1418		80	Bari	2.1106
55	Siena	1.1595		71	Chieti	2.2537
83	Lecce	1.1643		87	Crotone	2.2577
5	Novara	1.1787		88	Catanzaro	2.2841
17	Pavia	1.1938		7	Asti	2.3848
44	Ravenna	1.2138		64	Rieti	2.7114
24	Vicenza	1.2294		86	Cosenza	2.7446
36	Genova	1.2401		6	Cuneo	2.7882
92	Palermo	1.2710		8	Alessandria	2.8115
79	Foggia	1.2856		90	Reggio di Calabria	3.1382
31	Udine	1.2977				

Table 9.4   DEA efficiency results (A&P model), 2001

	Province	Score			Province	Score
46	Rimini	0.2740		60	Ancona	1.2271
28	Padova	0.3839		33	Trieste	1.2304
95	Caltanissetta	0.3882		36	Genova	1.2375
72	Isernia	0.3946		99	Siracusa	1.2392
21	Bolzano-Bozen	0.4400		97	Catania	1.2532
29	Rovigo	0.4831		16	Brescia	1.2833
27	Venezia	0.5852		103	Cagliari	1.3145
32	Gorizia	0.6007		68	L'Aquila	1.3333
4	Verbano-Cusio-Ossola	0.6322		12	Lecco	1.3488
14	Milano	0.6424		59	Pesaro e Urbino	1.3665
26	Treviso	0.6525		15	Bergamo	1.3698
25	Belluno	0.6989		81	Taranto	1.3794
35	Savona	0.7092		94	Agrigento	1.3812
50	Firenze	0.7117		96	Enna	1.4012
23	Verona	0.7360		24	Vicenza	1.4042
69	Teramo	0.7549		5	Novara	1.4059
70	Pescara	0.7990		91	Trapani	1.4124
22	Trento	0.8033		13	Sondrio	1.4190
51	Prato	0.8460		37	La Spezia	1.4279
78	Salerno	0.8691		101	Nuoro	1.4739
61	Macerata	0.8766		73	Campobasso	1.4844
66	Latina	0.9140		41	Modena	1.5093
62	Ascoli Piceno	0.9234		85	Matera	1.5161
52	Livorno	0.9340		75	Benevento	1.5164
34	Imperia	0.9462		100	Sassari	1.5191
42	Bologna	0.9570		48	Lucca	1.5295
83	Lecce	0.9759		67	Frosinone	1.5299
56	Grosseto	0.9875		79	Foggia	1.5650
39	Parma	0.9924		80	Bari	1.5702
53	Pisa	0.9964		10	Varese	1.6026
20	Mantova	1.0119		54	Arezzo	1.6505
47	Massa-Carrara	1.0368		3	Biella	1.6616
55	Siena	1.0416		71	Chieti	1.6983
65	Roma	1.0644		1	Torino	1.7367
11	Como	1.0645		74	Caserta	1.7435
49	Pistoia	1.0810		64	Rieti	1.7556
58	Terni	1.0880		98	Ragusa	1.7598
76	Napoli	1.0913		38	Piacenza	1.7949
17	Pavia	1.0926		82	Brindisi	1.8129
40	Reggio nell'Emilia	1.0960		2	Vercelli	1.9752
30	Pordenone	1.0979		84	Potenza	1.9948
9	Aosta	1.1095		8	Alessandria	2.0719
43	Ferrara	1.1212		86	Cosenza	2.2068
102	Oristano	1.1399		88	Catanzaro	2.5413
45	Forli'-Cesena	1.1466		6	Cuneo	2.7833
44	Ravenna	1.1493		89	Vibo Valentia	2.9235
19	Cremona	1.1523		7	Asti	3.2089
92	Palermo	1.1789		87	Crotone	3.4543
93	Messina	1.2005		90	Reggio di Calabria	3.8029
57	Perugia	1.2119		63	Viterbo	4.2216
31	Udine	1.2175				

In 2001, the efficiency coefficient changed for the provinces of Parma, Bologna and Lecce, when only the province of Lodi appears in their peer group, with a weight equal, respectively, to 0.79, 0.79 and 0.66. Even in this case, the share of weight λ_{18} cannot be ignored. According to the A&P model, the super-efficiency score appeared to decrease for the provinces of Milan, Treviso, Rimini, Prato, Pescara and Salerno (Table 9.4).

As a final part of our DEA application, the analysis focussed on the relationship between bed-nights market share[6] – often used to evaluate and rank the performance of destinations – and efficiency score (Figures 9.1 and 9.2). The comparison between bed-nights market share and efficiency score shows that a high efficiency score does not necessarily correspond with a high bed-nights market share. The scatter-plot graph for both years (Figures 9.1 and 9.2) shows a high concentration of efficient provinces on the top-left, indicating that the majority of efficient Italian provinces do not represent a substantial share of the total national bed-nights. This can probably be explained by the absence of a tourist function for these provinces (as we mentioned earlier). Most of the more efficient provinces are business areas. Tourist flows are low in absolute terms but high in comparison with their inputs and their key competitors. These provinces achieve efficiency by having the right balance between inputs and outputs.

The scatter-plot graph of the bed-nights market share and efficiency score, for both years, allows us to subdivide the provinces into three clusters (Figures 9.1 and 9.2):

1. the first, on the top-right, with a high efficiency level and high market share;
2. the second, on the top-left, with a high efficiency level and a low market share;
3. the third, on the bottom-left, with a low efficiency level and a low market share.

At first sight (Figures 9.1 and 9.2), a loss of competitiveness is evident: many provinces that in 1998 were in the second cluster appear to move to the third cluster in 2001, probably due to an insufficient increase of bed-nights compared with utilized inputs.

Among the best practice, we find in the first cluster, in both years, provinces such as Bolzano-Bozen (BZ), Venice (VE) and Rimini (RN). But, for example, Naples (NA) and Rome (RM) show in 2001, in comparison with 1998, an underuse of productive capability in relation to their tourist resources and 'key competitors' (Figure 9.2). So, they maintain almost the same bed-nights market share and efficiency score in both years.

---

[6] The bed-night market share is calculated as the total provincial bed nights (international and national) divided by total national bed-nights (international and national).

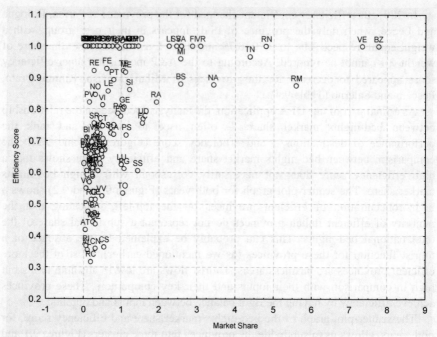

Figure 9.1   Efficiency score and bed-nights market share, 1998

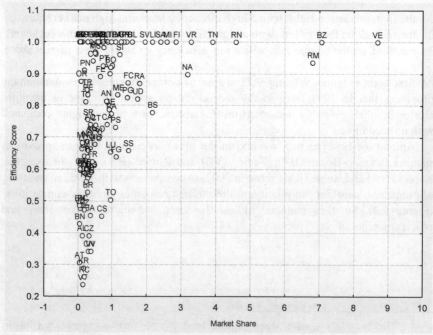

Figure 9.2   Efficiency score and bed-nights market share, 2001

In the second cluster there are typical tourist destinations (Trento (TN), Verona (VR), Ravenna (RA), Florence (FI), Siena (SI), Pisa (PI), Salerno (SA), Messina (ME), Palermo (PA), Siracusa (SR)) and business areas (i.e. Milan (MI), Padua (PD), Genova (GE), Livorno (LI), etc.). Finally, the third cluster, is largely made up of destinations without a particular tourist image and vocation.

9.5 Conclusions

In recent years, destination tourist performance, destination competitiveness and its measurement have increasingly been written about in the tourism literature. In fact, over the last few years, tourism has become an important activity with positive economic effects, because it is currently the fashion to mentally escape and relax by finding new, unusual and remote places to visit.

All this has prompted DMOs to plan and develop tourism policies based on strategies and operating actions that create an advantage over their competitors. As a consequence, it has become very important to measure the performance of each area against its 'key competitors' in order to identify the correct strategic actions needed to maintain or strengthen its position as market leader.

In spite of these needs, only a little empirical research – based on micro and qualitative data – is available in the literature (e.g. Kozak and Rimmington, 1999; Kozak, 2002), while an example of Destination Competitiveness Analysis by quantitative data is supplied by Alavi and Yasin (2000). In their papers, the authors used the shift-share technique to decompose the growth in tourist arrivals to four countries in the Middle East from six different regions of the world. The aim of the authors was to supply policy makers in the Middle East with a systematic approach to tourism policy. Our analysis belongs to this stream of the literature, and we present the results concerning the possibility of evaluating the performance and measuring the efficiency of Italian tourist destinations. It is an evaluation analysis of tourist performance by quantitative and macro-data.

For our purpose, Data Envelopment Analysis (DEA) was applied in order to evaluate the tourist efficiency of different provinces of Italy. We applied the CCR output-oriented model (Charnes et al., 1978) to the 103 Italian provinces, for the years 1998 and 2001, using two outputs and five inputs, and then we applied the A&P model (Andersen and Petersen, 1993) to rank the efficient destinations and to explore some possible outlier destinations.

For the years 1998 and 2001, we observed that there were 32 and 29 efficient provinces, respectively, and the average efficiency scores were 0.78 and 0.76, respectively. Among the better performers, which maintained their position over time, were provinces like Bolzano-Bozen (BZ), Verona (VR), Venice (VE), Rimini (RN) and Salerno (SA), all examples of cultural and scenic Italy.

The analysis showed that the inefficiency of many provinces is attributable to an imbalance between inputs and outputs. In particular, for traditional tourist destinations this can be interpreted as an underuse of their productive capability in relation to their tourist resources due to an inability to manage resources or as an expression of the phase of maturity of the tourist life cycle of the Italian product.

With DEA, we have found a suitable way of combining different indicators on the 'supply side' to explain how the provinces transform their resources into tourist flows. It follows that the best way to manage the various indicators is to insert them into a well-defined theoretical background (i.e. the production function). Although many factors influence the levels and changes in tourist supply, we have underlined, using the available data, some relevant features of the production process of this particular 'product'.

These results could make this field worth exploring further especially in order to compare this approach, which we have defined 'macro', with a micro-approach based on sample surveys. It would be a research challenge to explore the possibility of translating and putting into operation Sen's concept of *well-being* in tourism, following the specific theoretical perspective of the 'capability approach' (Sen, 1993). In brief, after the tourist has chosen his/her holiday destination, on the basis of his/her income and leisure time, then the achievable *holiday well-being* can be viewed as a function of the available combined tourist commodities (natural and cultural resources; amount and quality of accommodation and restaurants; accessibility to transportation systems; all the activities available at the destination and what the tourist-consumer will do during the visit; tourist safety; resident behaviour; past vacation experiences, etc.) that define the alternative combinations of *functionings*: total leisure experience, mental escape and relaxation, pleasure in unrepeatable experiences, body well-being, etc. (i.e. *capabilities*[7]).

Our goal would still be to develop a strict measurement system for *holiday tourist well-being*, but then it is important to focus on the complementary nature and importance of qualitative and quantitative performance measurements in defining a strategic tourism policy.

References

Alavi, J. and M.M. Yasin (2000), 'A Systematic Approach to Tourism Policy', *Journal of Business Research*, 48: 147–56.

Andersen, P. and N. Petersen (1993), 'A Procedure for Ranking Efficient Units in Data Envelopment Analysis', *Management Science*, 39(10): 1261–64.

Banker, R.D., A. Charnes and W.W. Cooper (1984), 'Some Models of Estimating Technical and Scale Inefficiencies in Data Envelopment Analysis', *Management Science*, 9(9): 1078–92.

---

[7]   Sen (1993) argues that 'well-being is best seen as an index of the person's functionings', where the *'functionings* represent part of the state of a person – in particular the various things that he or she manages to do or be in leading a life. The *capability* of a person reflects the alternative combinations of functionings the person can achieve, and from which he or she can choose one collection' (Sen, 1993, p. 31).

Buhalis, D. (2000), 'Marketing the Competitive Destination of the Future', *Tourism Management*, 21: 97–116.

Buhalis, D. and J. Fletcher (1992), *Environmental Impacts on Tourist Destinations*, Memorandum, University of Aegean, Mytilini.

Charnes, A., W.W. Cooper and S. Li (1989), 'Using Data Envelopment Analysis to Evaluate Efficiency in the Economic Performance of Chinese Cities', *Socio-Economic Planning Sciences*, 23(6): 325–44.

Charnes, A., W.W. Cooper and E. Rhodes (1978), 'Measuring the Efficiency of Decision Marking Units', *European Journal of Operational Research*, 2(6): 429–44.

Coccossis, H. and P. Nijkamp (eds) (1995), *Sustainable Tourism Development*, Ashgate, Aldershot, UK.

Crouch, G.I. and J.R.B. Ritchie (1999), 'Tourism, Competitiveness, and Societal Prosperity', *Journal of Business Research*, 44: 137–52.

Crouch, G.I.. and J.R.B. Ritchie (2000), 'The Competitive Destination: A Sustainability Perspective', *Tourism Management*, 21: 1–7.

Cuffaro, M. and E. Vassallo (2002), 'Sviluppo Economico e Sviluppo Umano: una Nota sulla Classificazione ONU di Alcuni Paesi', *Scritti di Statistica Economica*, 10, Liguori, Naples.

Farrell, M.J. (1957), 'The Measurement of Productive Efficiency', *Journal of the Royal Statistical Society*, 120: 211–81.

Fayos-Sola, E. (1996), 'Tourism Policy: A Midsummer Night's Dream?', *Tourism Management*, 17(6): 405–12.

Kozak, M. (2002), 'Destination Benchmarking', *Annals of Tourism Research*, 29(2): 497–519.

Kozak, M. and M. Rimmington (1999), 'Measuring Tourist Destination Competitiveness: Conceptual Considerations and Empirical Findings', *Hospitality Management*, 18: 273–83.

Macmillan, W.D. (1986), 'The Estimation and Applications of Multi–regional Economic Planning Models Using Data Envelopment Analysis', *Papers of the Regional Science Association*, 60: 41–57.

Martić, M. and G. Savić (2001), 'An Application of DEA for Comparative Analysis and Ranking of Regions in Serbia with Regard to Social-Economic Development', *European Journal of Operational Research*, 132: 343–56.

Newall, J.E. (1992), 'The Challenge of Competitiveness', *Business Quarterly*, 56.

OECD (1994), *The World Competitiveness Report*, World Economic Forum and IMD International, Lausanne, Switzerland.

Poon, A. (1993), *Tourism, Technology and Competitive Strategies*, CAB, Oxford.

Porter, M.E. (1990), *The Competitive Advantage of Nations*, The Free Press, New York.

Prosser, R. (1994), 'Societal Change and Growth', in E. Carter and G. Lowman (eds), *Alternative Tourism, Ecotourism a Sustainable Option?*, John Wiley & Sons, Chichester, pp. 89–107.

Scheel, H. (2000), *Efficiency Measurement System*, version 1.3.

Scott, B.R. and G.C. Lodge (1985), *U.S. Competitiveness in the World Economy*, Harvard Business School Press, Boston.

Sen, A. (1993), 'Capability and Well-Being', in M. Nussbaum and A. Sen (eds), *The Quality of Life*, Clarendon Press, Oxford, pp. 30–53.

152   *Tourism and Regional Development: New Pathways*

Susiluoto, I. and H. Loikaanen (2001), 'The Economic Efficiency of Finnish Regions 1988–1999: An Application of the DEA Method', Paper presented at the *41st Congress of the European Regional Science Association*, 29 August-1 September 2001, Zagreb, Croatia.

Chapter 10

Delineating Ecoregions for Tourism Development

Thomas Hatzichristos, Maria Giaoutzi and
John C. Mourmouris

10.1 Introduction

Over the past 50 years tourism has become a major activity in our society and an increasingly important sector in terms of economic development, mainly as a result of income increase and improvements in transport systems. Tourism though is not only a rapidly rising economic activity, on all continents, in both countries and regions, but also a growth sector which may have increasingly many adverse effects on environmental quality (Giaoutzi and Nijkamp, 1994).

Meanwhile an increasing political interest has developed regarding the disruption of the earth's natural resources and environmental decay. Policymaking bodies believe that – despite the potential catastrophes incorporated in our modern way of life – human resources, knowledge and capabilities are available to create a sustainable development. *Sustainable development* has become a popular concept in this respect, defined as paths of human progress which meet the needs and aspirations of the present generation without compromising the ability of future generations to meet their needs (Giaoutzi and Nijkamp, 1994).

In the context of the worldwide debate on sustainable development, there is also an increasing need for a thorough reflection on sustainable tourism, where the socio-economic interests of the tourist sector are brought into harmony with environmental constraints, now and in the future.

Tourism is intricately involved with environmental quality, as it directly affects the natural and human resources and at the same time is conditioned by the quality of the environment.

Such a relationship has important implications from the point of view of policies, management and planning. Examples of such policies may include the creation of a flexible legal framework, the provision of transport facilities, the supply of support services, the establishment of land use zoning for tourist purposes, the provision of financial and fiscal incentives, the design of marketing tools, etc.

The main aspect in the context of sustainable tourism is environmental quality and therefore qualitative issues should be introduced rather systematically at all levels of the planning process.

The re-assessment and development of appropriate tools for policymaking dealing with qualitative aspects in the systems under consideration greatly supports management and planning for sustainable tourism.

Methods of land use zoning and the delineation of regions for tourist purposes are important tools in this context. Their delineation involves deducing or eliminating details over large areas. The basis for discussion in the present chapter is what is called the 'ecoregion' or environmental region, a type of region used in describing environmental aspects and resources. For the delineation of these regions, several methods have been proposed which have a broad range of limitations (see Carver, 1992; Skidmore, 1989; Leung and Leung, 1993; Moore et al., 1991; Brown et al., 1998).

In order to overcome such limitations, the present chapter focusses on the development of a methodology for the delineation of ecoregions utilizing fuzzy logic and Geographic Information Systems (GIS). GIS provides a powerful set of tools for the input, management and presentation of data. Fuzzy logic, on the other hand, enables the treatment of environmental phenomena, which are not exact or precise but rather fuzzy. The integration of fuzzy logic in a GIS context increases the analytical capabilities of the proposed methodology which deals with some of the limitations met in the previous approaches.

10.2 The use of the ecoregion concept in planning

The adoption of the appropriate concepts in designing methodologies is of great importance in planning and policy making since this greatly enhances the effectiveness of the policy efforts. In the planning context for tourism development, efforts have been undertaken to identify methods and techniques which comply with the goal of sustainable development, and more precisely with the goal of sustainable tourism development (Giaoutzi and Nijkamp, 1994). The concept of 'ecoregion' has emerged out of this need. Therefore, in the present chapter, the ecoregion will be used, as most complying with the sustainability concept, in the methodological framework developed below.

The term 'ecoregion' was originally used by Crowley (1967) and first applied by Bailey (1976) in mapping the ecological regions of the USA. Ecoregions are nowadays generally accepted as regions of relative homogeneity in terms of ecological characteristics or relationships between organisms and their environment and therefore constitute a valuable tool for the management of natural resources as well as for environmental impact assessment (Omernik et al., 1989).

To date, several qualitative and quantitative methods have been proposed for the delineation of ecoregions. Some researchers, for example, Ceballos and Lopez (2003) suggest multicriteria analysis for the delineation of suitable areas for crops;

Skidmore (1989) uses expert systems for the classification of forest types, while Leung and Leung (1993) use the same method for ecological modelling purposes. Moore et al. (1991), on the other hand, utilized decision trees for the prediction of vegetation distribution. Finally, Brown et al. (1998) suggest supervised classification using maximum likelihood or neural networks for the classification of glaciated landscapes, while other researchers suggest combinations of the above. The main drawback of the above methodologies is the strict nature of their output that does not reflect the fuzziness inherent in reality, since 100 per cent certainty in defining environmental indicators/criteria used for the delineation of regional boundaries is not feasible.

10.3 Methodological approach

The proposed approach of land use zoning and delineation of regions for sustainable tourist development is based on the integration of fuzzy logic in a GIS context. It aims at overcoming limitations of previous approaches in this respect (Carver, 1992; Skidmore, 1989; Leung and Leung, 1993; Moore et al., 1991; Brown et al., 1998), in order to create more policy relevant regions for sustainable tourism.

10.3.1 Using GIS for the delineation of ecoregions

The GIS approach as a box of tools for handling geographical information is rather important, since it provides a list of impressive options. In most GIS packages, though functionality, in spatial analysis, lies mainly with the ability to perform deterministic overlay and buffer functions (Carver, 1992). Such abilities, whilst ideal for performing spatial searches based on nominally mapped criteria, have certain limitations when multiple criteria and targets are applied, as in the case of delineating ecoregions. The integration of GIS with analytical techniques will be a valuable addition in the GIS toolbox. According to Fotheringham and Rogerson (1994), progress in this area is inevitable and future developments will continue to place increasing emphasis upon the analytical capabilities of GIS.

A tool for increasing the analytical capabilities of GIS, in the present context, is the integration of the fuzzy set approach in the GIS platform as most suited to applications where decision criteria are not rigid and the boundaries between two regions are not discreet. Inexact boundaries or class overlapping appear to be the rule, rather than the exception, in geographical problems (Openshaw, 1997). The next subsection provides a brief presentation of the fuzzy logic approach and its relevance for spatial analysis.

10.3.2 The fuzzy logic approach

The classic Boolean logic is binary, that is, a certain element is true or false, or an object belongs to a set or it does not. Fuzzy logic was introduced by Zadeh in 1965, and permits the notion of nuance. Apart from being true, a proposition may also be anything from almost true to hardly true (Kosko, 1991). In comparison with the Boolean sets, a fuzzy set does not have sharply defined boundaries. The notion of a fuzzy set provides a convenient way of dealing with problems in which the source of imprecision is the absence of sharply defined criteria of class membership rather than the presence of random variables.

As already mentioned above, a significant proposition in statistical logic is that each point of a set U is unequivocally grouped with the other members of its group and thus bears no similarity to members of other groups. One way to characterize an individual point's similarity to all the groups is to represent the similarity a point shares with each group with a function (termed the membership function), whose values (called memberships) are between $0 < m < 1$. Each point will have a membership in every group; memberships close to unity signify a high degree of similarity between the point and a group, while memberships close to zero imply little similarity between the point and that group. Additionally, the sum of the memberships for each point must be unity. The complement of A is NOT A.

Although in Boolean logic, A and NOT A are unique, in fuzzy logic the following equation is true:

$$m_{notA} = 1 - m_A$$

Fuzzy degrees are not the same as probability percentages. Probabilities measure whether something will occur or not. Fuzziness measures the degree to which something occurs or some condition exists. Crisp sets are a subset of fuzzy sets. Only when an object belongs 100 per cent to a group are fuzzy sets identical to crisp sets.

In order to solve a problem with a knowledge-based fuzzy system it is necessary to describe and process the influencing factors in fuzzy terms and provide the result of this processing in an easy to handle form. The basic elements of a knowledge-based fuzzy system are:

- fuzzification;
- knowledge base;
- processing;
- defuzzification.

These elements are described in detail in the following paragraphs. Several types of membership functions can be utilized (Burrough, 1996). The membership function reflects the knowledge for the specific object or event.

Every continuous mathematical function can be approximated by a fuzzy set. For example, the criterion 'distance from a road' can be approximated from the membership function illustrated in Figure 10.1.

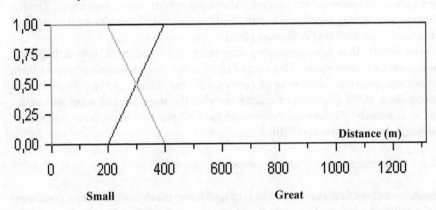

Figure 10.1 Membership function for the criterion 'distance from a road'

The assignment of a membership function to every variable of the problem is called a *fuzzification process*. During this process crisp subsets are transformed into linguistic subsets, such as 'small' or 'great' distance (Figure 10.1). The concept of the linguistic variable illustrates particularly clearly how fuzzy sets can form the bridge between linguistic expression and numerical information. The most widely used form of membership function is the triangular. Its maximum is in the most representative value. Other forms also exist, such as the trapezoid, the Gaussian, etc. For the determination of these functions, various methods have been utilized.

The *second step* in the fuzzy systems approach is the *definition of the rules* that connect the input with the output. These rules are based on the form 'if... then'. The knowledge in a problem-solving area can be represented by a number of rules. Experts with broad knowledge on the specific field usually undertake the task of rule definition. There is no need for assigning weights in the criteria used. The weights are implicitly taken into account through the rules defined. For example, if the output set 'suitability' is comprised of two subsets called 'poor' and 'appropriate', the rules could be:

- If the distance is short, then suitability is poor.
- If the distance is long, then suitability is appropriate.

The next step is the *processing of the rules*. This step is also called *inference*. It is comprised of three stages: aggregation; implication; and accumulation. *Aggregation* provides the degree of fulfilment for the entire rule concerned. All the Boolean algebra operations (like intersection, union, negation, etc.) can be easily extended to fuzzy set operations (Bezdek, 1981) and can be used in this stage. *Implication* determines the degree of fulfilment of the conclusion. Finally, *accumulation* brings together the individual results of the variables used. Details of this process can be found in Bezdek (1981).

The result of a rule processing can be transformed back into a linguistic expression or a crisp value. This second process is called defuzzification and there are several methods to achieve it (Leekwijck and Kerre, 1997). The simplest among them is the selection of the subset with the maximum membership value, e.g. *fuzzy results*: 73 per cent poor suitability; 37 per cent appropriate suitability; *defuzzified* results: poor suitability.

10.3.3 Integration of GIS with fuzzy logic

The delineation of ecoregions using GIS and fuzzy logic is a procedure consisting of the following stages (Figure 10.2). The environmental criteria to be used and the delineation procedure of ecoregions largely depend on the scope of the project and the objectives set during this first stage (Part A of Figure 10.2).

The next stage of the process is the definition of the geographic layers describing the natural resources of the study area and the impact of human activities (Part B of Figure 10.2). Geographical data and descriptions for the criteria that are perceived to affect the issues of concern are gathered during this stage. The source of the criteria (geographic layers) and their resolution is also decided. A quality control process is appropriate at this point. It is likely that additional or revised information will be needed later.

The geographic data are introduced in the GIS concerned and after appropriate handling they create the database upon which the delineation of ecoregions will be based. (Part C of Figure 10.2). It is also likely that additional or revised information will be used later. The more well-defined the objectives at this point the less will be the additional information used.

Moreover, before applying fuzzy logic for the delineation of ecoregions, the geographic database needs to be transformed into a raster format (Figure 10.3). Overlaying creates a new raster layer that combines the attributes of all previous layers used as an input into the fuzzy logic software (Part D of Figure 10.2).

The output of the fuzzy set analysis is a layer with membership values for every pixel in the pursued regions. The pixels of each ecoregion have a value ranging from 0 to 100 depending on their membership value in the final layer. This procedure is illustrated in Figure 10.3 for the three regions.

These final membership values are imported back into a GIS for illustration purposes (Part E of Figures 10.2 and Figure 10.3).

In the following section the methodology described above will be applied in a specific case context, in order to demonstrate the advantages of the approach.

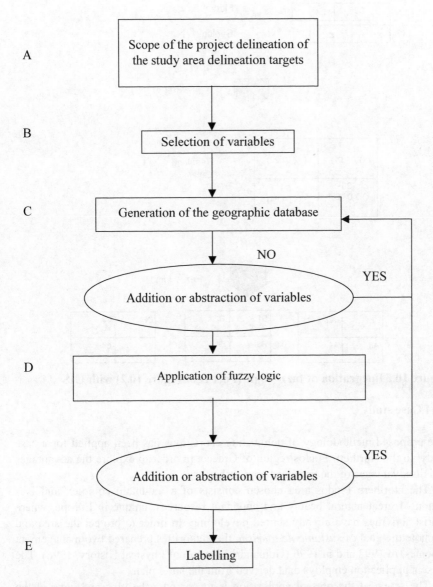

Figure 10.2 Methodological framework of ecoregion delineation

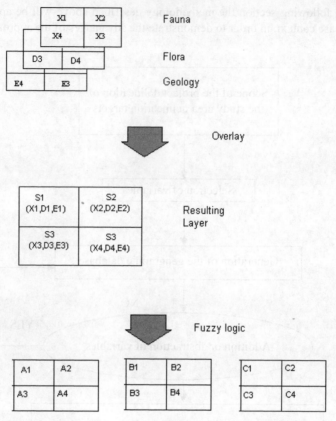

Figure 10.3 Integration of fuzzy logic (Part D of Figure 10.2) with GIS

10.4 Case study

The proposed methodology of delineating ecoregions has been applied for a case study, in the Northern Pindos region of Greece, in order to explore the advantages and disadvantages of the proposed approach.

The Northern Pindos area chosen consists of a mountain complex and is a region of great natural beauty and abundant resources, unique in Europe, where tourist activities have already started developing. In order to protect the area and promote the goal of *sustainable tourism*, the authorities prepared two management schemes, in 1992 and in 1996 (Goulandri Museum of Physical History, 1996). The present application employs data derived from the latter plan.

The scope of the present application is to compare the pros and cons of the proposed methodology versus the previous methodologies applied. A number of characteristics, both quantitative and qualitative will be presented in order to introduce into the delineation process aspects that better describe the ecoregion concept, as a tool for pursuing sustainable tourism.

10.4.1 Description of the study area

The study area is located on the north-west borders of Greece, in the regions of Epirus and Western Macedonia, between latitudes 39 41' and 40 15' north-south and longitudes 20 35' and 21 20' east-west, just 10 km from the Albanian border (Map 10.1). The area includes the mountains of Timfi, Lygkos, Smolika and Orliaka, which are part of the northern part of the Pindos mountain complex.

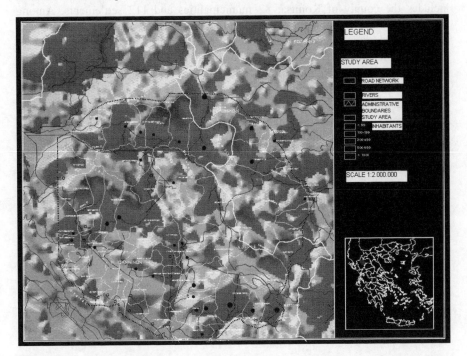

Map 10.1   Area studied

10.4.2 Relief

The area studied is crossed by mountain ranges oriented roughly North-West/South-East. Opposite the city of Ioannina, in the northern part of the area studied, rises the grey wall of Mt Mitsikeli. This mountain is uniform in height (about 1800m) up to its south-eastern edge, whereupon it suddenly dips, to form a saddle of 1000m in height. This saddle constitutes the border between the Northern and Southern Pindos. Eastwards, sweeping across the horizon are forest-covered compact mountain masses, whose peaks are deep red in colour. These are Mt Lakmon and Mt Timfi, 900 and 1500m high.

A panoramic view of the region can be seen from the crest of Mt Mavrovouni in the west. Viewed from this point, the mountains reveal their true shape, forming a broad, mountainous whole, rather than a single crest. They are flanked by gorges containing the tributaries of the River Aoos on one side and of the River Aliakmon on the other. The area is covered in forest and is the wildest part of the Pindos.

10.4.3 Administrative organization of the area

The area studied mainly consists of the Prefectures of Ioannina and of Grevena. It includes the county of Konitsa, 82 municipalities and 112 settlements. Among these, 66 of the municipalities are located in the Prefecture of Ioannina; 15 Communities are situated in the Province of Grevena, and the Settlement of Eptahori is located in the Province of Kastoria. The cities of Ioannina, Grevena, Metsovo and Konitsa are the administrative centres of their respective Prefectures and Provinces. All the administrative and social functions of the area are concentrated in the cities. The population of the cities are, 45,000, 8,000, 2,700 and 3,000, respectively. The area studied covers approximately 229,500ha, about 60 x 50km, which represents about 30 per cent of the total Prefecture area. The largest Community is Perivoli, in the Prefecture of Grevena, 137sq km in area, whilst the smallest is Molista, of 4.8sq km. The average is 27sq km.

The area is mountainous with a high amplitude of relief. There is a large percentage of mountain Communities, 79 per cent and 47 per cent, respectively, for the Prefectures of Ioannina and Grevena. As regards the area studied, the settlements are located at levels between 400 and 1,450m. Flat areas are rare and used for cultivation.

10.4.4 Geographical database

The analysis of the natural environment employs the following layers and variables:

- Topography;
- Geology;
- Soil suitability;
- Flora;
- Fauna;
- Land use.

The 'settlements' layer is also appropriate. These layers were digitized in vector format and are described briefly in the following paragraphs.

10.4.4.1 Topography
A digital terrain model was used on a scale of 1:250,000. GIS technology was employed, to generate slopes which are depicted in Map 10.2 below, and categorized into six classes. The values for slopes range from 0 to 100 per cent.

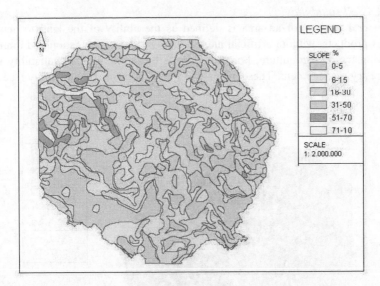

Map 10.2   Slopes in the study area

10.4.4.2 Geology
The map indicating the existence geological formations (see Map 10.3) in the study area was produced by the Greek National Geological Organization (scale 1:250,000).

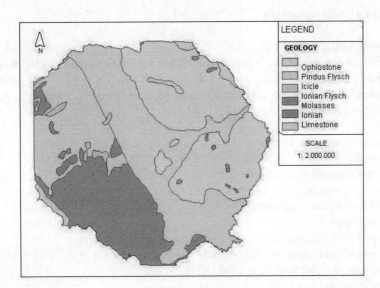

Map 10.3   Geology

10.4.4.3 Soil suitability

The natural fertility of an area is defined as the ability of the land to generate various products, without artificial aids such as fertilizers, irrigation and drainage. The Ministry of Agriculture, Forestry Service has compiled soil suitability maps for forestry management. These maps give five classes of soil suitability (see Map 10.4).

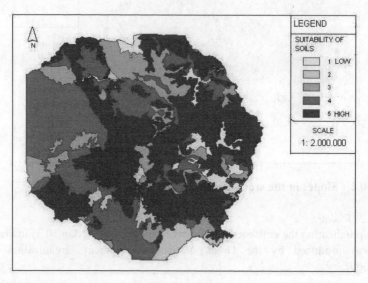

Map 10.4   Soil suitability

10.4.4.4 Flora

The natural features of the region were studied by the Goulandris Museum group (Goulandris Study, 1994). The study presents the following categories of ecosystems, which, when horizontally ordered, correspond to zones of naturally occurring vegetation (see Map 10.5):

1. Partially wooded, with fir.
2. Partially wooded with beech.
3. Forest of deciduous oaks.
4. Partially wooded, with deciduous oaks.
5. Fields and cultivated areas.
6. Barren area.
7. Forest of deciduous oak and partially wooded with black pine.
8. Forest of deciduous oak and partially wooded with beech.
9. Partially wooded with black pine and deciduous oaks.
10. Forest of deciduous oak and areas of restored fir.
11. Forest of deciduous oak and areas of restored black pine.

12. Forest of fir.
13. Partially wooded with black pine.
14. Forest of beech.
15. Areas of restored black pine.
16. Forest of black pine and deciduous oak.
17. Forest of beech and partially wooded with black pine.
18. Forest of beech and areas of restored fir.
19. Forest of deciduous oak, fir and black pine.
20. Forest of beech and fir.
21. Forest of deciduous oak, beech and black pine.
22. Forest of black pine and partially wooded with beech.
23. Forest of beech, fir and black pine.
24. Forest of black pine and areas of restored beech.
25. Restored areas of beech and forest of black pine.
26. Partially wooded with black pine and fir.
27. Partially wooded with black pine and beech.
28. Restored areas of forest of black pine.
29. Forest of black pine and partially wooded with fir.
30. Forest of fir and partially wooded with black pine.
31. Forest of black pine and areas of restored fir.
32. Forest of black pine.
33. Forest of deciduous broad-leafed trees.
34. Partially wooded with deciduous broad-leafed trees.
35. Bare grasslands and meadows.
36. Evergreen broad-leafed trees.
37. Evergreen broad-leafed trees and partially wooded with fir.
38. Forest of deciduous oak and deciduous broad-leafed trees.
39. Forest of deciduous oak and partially wooded with broad-leafed trees.
40. Forest of deciduous broad-leafed trees and partially wooded with black pine.
41. Forest of beech and partially wooded with deciduous broad-leafed trees.
42. Restored areas of forest of fir in bare grasslands and meadows.
43. Forest of white pine.
44. Partially wooded with white pine.
45. Arkefthos.
46. Evergreen broad-leafed trees and partially wooded with white pine.

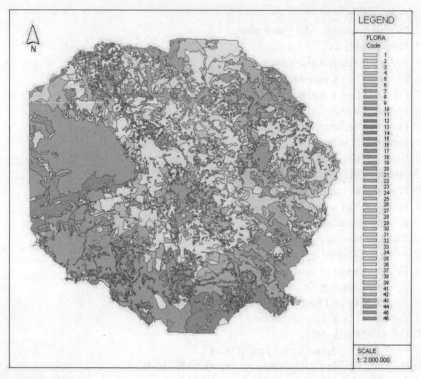

Map 10.5   Flora

10.4.4.5 Fauna

The fauna consists of mammals, reptiles and birds. The isolation, the limited access, and the variety of ecosystems of the Pindos mountain complex provide an ideal environment – unique for both Greece and the rest of Europe – which enables species that have died out in other parts of the country still to be found here. These species have been mapped, by use of a digitized model of the region, by specialist biologists who also included the flora in their investigations, at a scale of 1:100,000.

On the basis of the ecological value of the fauna which was defined on the basis of a weighting procedure, the variable *Fauna* takes values from 0 to 525, on a hierarchical scale, as illustrated in Map 10.6.

10.4.4.6 Land use

The types and coding of land uses in the area are derived from the elements employed in the land cover *Corine* programme (scale 1:100,000). The body responsible for *Corine* in Greece was the Greek Cadastral and Cartographic Organization. The land cover types in the study area are depicted in Table 10.1 and Map 10.7.

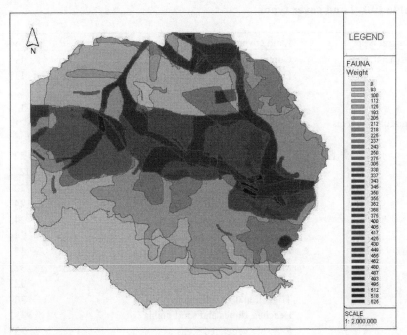

Map 10.6   Fauna

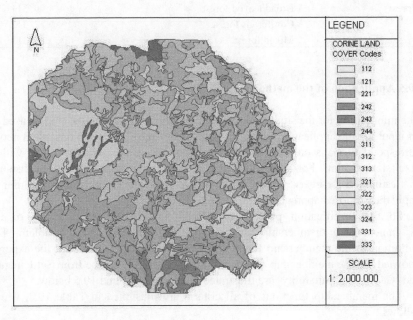

Map 10.7   Land cover types
Note: See Table 10.1 for explanation of codes.

Table 10.1 Land cover types

General types	Land cover types	Code
Artificial areas	Continuous urban fabric	111
	Discontinuous urban fabric	112
	Industrial or Commercial units	121
	Airports	124
Agricultural areas	Non irrigated arable land	211
	Continously irrigated arable land	212
	Vineyards	221
	Fruit trees and berry plantations	222
	Olive grooves	223
	Complex cultivations	242
	Land principally occupied by agriculture	243
	Agro-forestry areas	244
Bushy areas	Natural grassland	321
	Moors and Heath land	322
	Sclerophyllous vegetation	323
	Transitional woodland shrub	324
	Beaches, dunes and sand plains	331
	Sparsely vegetated areas	333
	Burnt areas	334
Forest	Broad leaved forest	311
	Coniferous forest	312
	Mixed forest	313

10.5 Application of the method

The information for the study area presented in the previous section will be used as an input for the application of the proposed approach. The layers described above correspond to maps on a scale 1:200,000, converted into a raster format with a pixel size of $50m^2$. For all layers that had to be buffered, the Euclidean distance was estimated for every pixel. The final layers were overlaid in one, in order to apply the proposed approach.

So the fuzzification process introduced at the *first stage* refers to the assignment of a membership function to every variable of the problem. The definition of the membership function for every variable is based on the expert's knowledge. The membership functions for the criteria 'Distance from Settlements' and 'Geological Suitability' are illustrated in Figures 10.4 and 10.5 below.

The membership functions of all criteria are illustrated in Table 10.2, which follows.

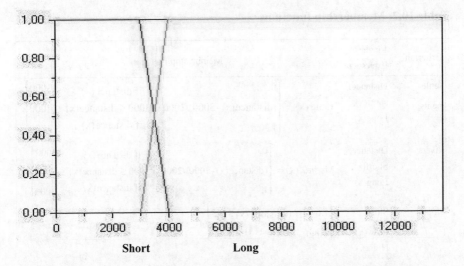

Figure 10.4 Membership function for 'Distance from Settlements'

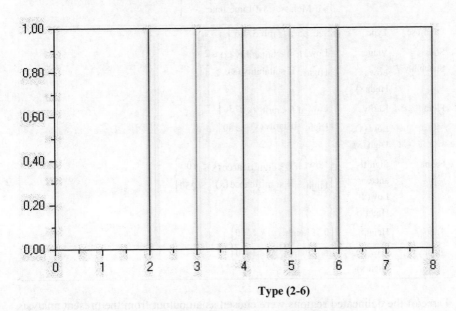

Figure 10.5 Membership function for 'Geological Suitability'

Table 10.2 Membership functions

Criterion	Linguistic Expression	Membership Function
Settle-ments	Distance: short(x), Long(x)	$\text{Long}(x) = \begin{cases} 0, & \text{if distance}(x) < 3000 \\ (\text{distance}(x) - 3000)/1000, & \text{if } 3000 \le \text{distance}(x) \le 4000 \\ 1, & \text{if distance}(x) > 4000 \end{cases}$
Rivers	Distance: Short(x), Long(x)	$\text{Μεεγαλ}(x) = \begin{cases} 0, & \text{if distance}(x) < 1000 \\ (\text{distance}(x) - 1000)/2000, & \text{if } 1000 \le \text{distance}(x) \le 3000 \\ 1, & \text{if distance}(x) > 3000 \end{cases}$
Slopes	Slope: Low (x), High(x)	$\begin{cases} \text{Low, if } 0 < \text{slope}(x) \le 3.5 \\ \text{High, if slope}(x) \ge 4 \end{cases}$
Geology	Type	$\begin{cases} 2 \text{ if ophiostone; } 3 \text{ if Pindus flysch;} \\ 4 \text{ if icicle; } 5 \text{ if Ionian flysch;} \\ 6 \text{ if Molasses; } 8 \text{ if Limestone.} \end{cases}$
Land-use	Type	$\{\text{See paragraph } 3.4.6.\}$
Soils Suitability	Value: Low (x), High(x)	$\begin{cases} \text{Low, if } 0 < \text{suitability}(x) \le 2 \\ \text{High, if suitability}(x) \ge 4 \end{cases}$
Flora	Rarity : Low (x), High (x)	$\begin{cases} \text{Low, if } 0 < \text{rarity}(x) \le 30 \\ \text{High, if rarity}(x) \ge 40 \end{cases}$
Fauna	Signifi-ance: Low(x), High(x)	$\begin{cases} \text{Low, if } 0 < \text{significance}(x) \le 100 \\ \text{High, if significance}(x) > 250 \end{cases}$
Topo-graphy	Height: Low (x), High(x)	$\begin{cases} 0, \text{if height}(x) < 1600 \\ 1, \text{if height}(x) \ge 1600 \end{cases}$

Three of the delineated regions were chosen as an output from the present analysis so that they could be compared with the results of the environmental management plan of 1996, which proposed three protected zones in the study area.

The *second step* in the proposed approach is the definition of the rules linking the input with the output. These rules are based on the form 'if... and..., then'. Knowledge in a problem-solving area can be represented by a number of rules. In the specific case study the following twelve rules were formulated:

1. If the Significance of Fauna is high, then the area falls within Ecoregion 1 with 85 per cent certainty.
2. If the Rarity of Flora is high, then the area falls within Ecoregion 1 with 65 per cent certainty.
3. If the Distance from Settlements is short, then the area falls within Ecoregion 3 with 70 per cent certainty.
4. If Slope is High and distance from rivers is short, then the area falls within Ecoregion 1 with 80 per cent certainty.
5. If Height is High and Slope is High, then the area falls within Ecoregion 1 with 85 per cent certainty.
6. If Geological Suitability is High and Height is High, then Ecoregion 1 with 65 per cent certainty.
7. If Soil Suitability is High and Geological Type is Ionian Limestone, then the presence of Ecoregion 1 is with 70 per cent certainty.
8. If Land-use is Brush land, then the presence of Ecoregion 1 is with 60 per cent certainty.
9. If Soil Suitability is High and Geological Type is Pindus Flysch or Ophiostone, then the presence of Ecoregion 2 is with 80 per cent certainty.
10. If Land-use type is Coniferous forest, or Mixed forest or transitional areas, then the area falls within Ecoregion 2 with 75 per cent certainty.
11. If the Significance of Fauna is Low and Land-use type is broad-leaved forest or agricultural land, then the area falls within Ecoregion 3 with 70 per cent certainty.
12. If Geological type is Ionian Flysch or Pindus Flysch and the Rarity of Flora is Low, then the area falls within Ecoregion 3 with 60 per cent certainty.

The *third step* of the process is the processing of the rules, called also inference. It is comprised of three stages: aggregation; implication; and accumulation. The operators used were minimum, minimum, and algebraic product, accordingly.

The results of the processing of the rules are illustrated in the three following maps. Ecoregion 1 is presented in Map 10.8, Ecoregion 2 is presented in Map 10.9, while Ecoregion 3 is presented in Map 10.10. Pixels with membership values close to unity signify areas with high membership values, while pixels with values close to zero imply areas with low membership values.

These results can be transformed back into a crisp map. This process is called defuzzification and is carried out by the selection of the Ecoregions with the maximum membership value. The results are illustrated in Map 10.11.

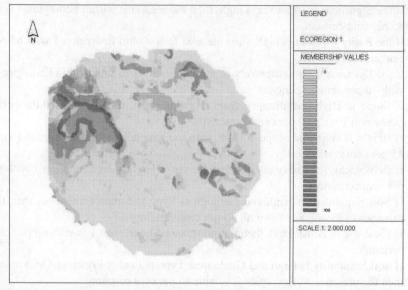

Map 10.8   Membership values of Ecoregion 1

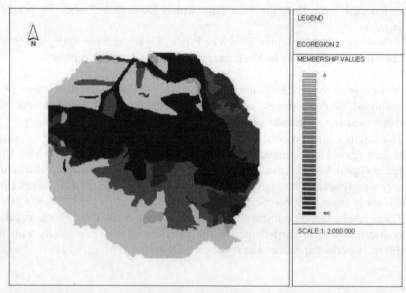

Map 10.9   Membership values of Ecoregion 2

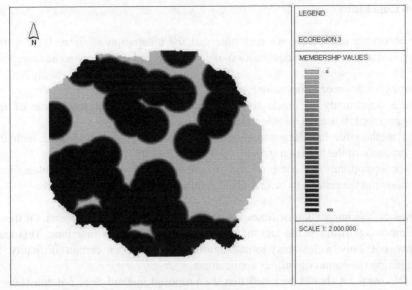

Map 10.10 Membership values of Ecoregion 3

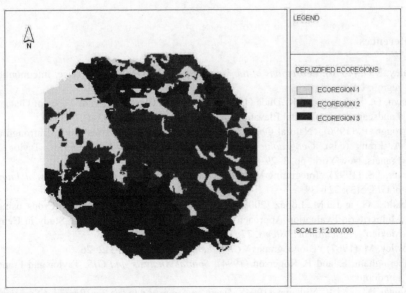

Map 10.11 Defuzzified Ecoregions

10.6 Conclusions

By observing the results, we may note that the integration of fuzzy logic in the proposed approach for the delineation of ecoregions has the following advantages:

- maximization of information;
- the opportunity to create a continuous classification of the whole of the geographical area in question, for all the regions;
- a method for handling and processing information that resembles with the rationale of the human mind;
- the opportunity to combine the parameters involved in the problem, thus allowing the criteria to be processed simultaneously.

However, the integration of fuzzy logic in GIS involves some problems. Of these, the most significant is the lack of ready-to-use membership functions. This may imply not only a less-easy-to-use approach but also a certain difficulty in comparing the output of various applications.

Of course, it should be noted that the proposed method does not constitute a specialized approach just for delineating ecoregions for tourism development. Its application should rather be extended to delineate other types of regions, e.g. demographic.

References

Bailey, R.G. (1976), *Ecoregions of the United States*, USDA Forest Service, Intermountain region, USA.

Brown, D., D. Lusch and K. Duda (1998), 'Supervised Classification of Types of Glaciated Landscapes using Digital Elevation Data', *Geomorphology*, 21(3-4): 233–50.

Burrough, P. (1996), 'Natural Objects with Indeterminate Boundaries', in P. Burrough and A. Frank (eds), *Geographic Objects with Indeterminated Boundaries*, Taylor and Francis, New York, pp. 3–29.

Carver, J.S. (1992), 'Integrating Multi-Criteria Evaluation with GIS', *International Journal of GIS*, 5(3): 321–39.

Ceballos, G. and J.M. Lopez (2003), 'Delineation of Suitable Areas for Crops using a Multicriteria Evaluation Approach and Land Use Mapping in a Case Study in Central Mexico', *Agricultural Systems*, 77: 117-136.

Crowley, M. (1967), 'Biogeography', *Canadian Geography*, 11: 312–26.

Fotheringham, S. and P. Rogerson (1994), *Spatial Analysis and GIS*, Taylor and Francis, London.

Giaoutzi, M. and P. Nijkamp (1994), *Decision Support Models for Regional Sustainable Development*, Avebury, London.

Goulandri Museum of Physical History (1996), *Management Plan for the Mountainous Area of Northern Pindos*, Management Plan Report No. 1253 (in Greek), Ministry of the Environment, Athens, Greece.

Kosko, B. (1991), *Neural Networks and Fuzzy Systems*, Prentice Hall, New York.

Hart, J.F. (1982), 'The Highest Form of the Geographer's Art', *Annals of the Association of American Geographers*, 72: 1–29.

Leekwijck, W. and E. Kerre (1999), 'Defuzzification: Criteria and Classification', *Fuzzy Sets and Systems*, 108: 159–178.

Leung, Y. and K.S. Leung (1993), 'An Intelligent Expert System Shell for Knowledge-Based Geographical Information Systems', *International Journal of Geographical Information Systems*, 7: 189–99.

Moore, D.M., B.G. Lees and S.M. Davey (1991), 'A New Method for Predicting Vegetation Distributions using Decision Tree Analysis in a Geographic Information System', *Environmental Management*, 15: 59–71.

Omernik, J., A. Gallant, T. Whittier, D. Larsen and R. Hughes (1989), *Regionalization as a Tool for Managing Environmental Resources*, EPA /600/3-89/060, USA.

Openshaw, S. (1997), *Artificial Intelligence in Geography*, Wiley, London.

Skidmore, A.K. (1989), 'An Expert System Classifies Eucalypt Forest Types Using Landsat Thematic Mapper Data and a Digital Terrain Model', *Photogrammetric Engineering and Remote Sensing*, 55: 1449–64.

Zadeh, L.A. (1965), 'Fuzzy Sets', *Information and Control*, 8: 338–353.

Chapter 11

Tourism and the Political Agenda: Towards an Integrated Web-based Multicriteria Framework for Conflict Resolution

Andrea De Montis and Peter Nijkamp

11.1 Introduction

During the last 30 years, the successful development of various regions endowed with natural capital has evoked confidence in tourism as a catalyst for economic and cultural growth. In developing countries neo-classical economists have advocated policies for a higher level of tourist activities and revenues in the hope of obtaining overall higher performances in the economic system.

There have been many changes in this paradigm, though. Recent advances in studies on the concept of integrated tourist development (Pearce, 1989; Wall, 1997) have suggested that tourism per se cannot be seen as a guaranteed factor of development, since it has to be linked to other economic sectors. Furthermore, research findings (APDR, 2000) indicate that integrated tourist development involves many conflicting stakes and interests. Butler (2000) recalls that, in many studies on environmental policy and planning, integration of tourism with other economic sectors is often regarded as a strategic goal. Indeed, integration is frequently associated with sustainability, since the objectives of the development are generally linked to the human and material resources embedded in the local context. In this case, sustainable development can be supported by the attitude of local communities towards social learning and self-organizing.

The introduction of policies furthering tourist activities into the agenda of an institutional body is frequently connected to the rise of sharp political debates among the different parties involved. These political coalitions usually endorse different value systems, are often concerned with interests linked to important territorial investments, and represent future perspectives of a number of stakeholders.

In those delicate decision-making environments, multicriteria analysis has proven to be a useful tool for supporting planning processes, at least if it helps stakeholders in confronting their different value systems in order to reach a reasonable compromise solution. Acknowledging the presence of some subjectivity in using multicriteria decision analysis, the need to open up the framework traditionally set for multicriteria methods and to conceive of new strategies for involving local societies in a dynamic, adaptive and interactive pattern becomes apparent.

Against this background, the aim of this chapter is to present an experimentation with a multicriteria method, with the goal of involving as many stakeholders as possible from all social groupings in the actual decision-making process. In this case study, the method is used to measure the achievement in integrating tourism with the whole economic system, on the basis of a case study applied to a set of municipalities located along the coast of the province of Cagliari, Italy.

The contents are articulated as follows. In Section 11.2 next, an innovative economic model of tourism activities is described that could support environmental policy and planning. In Section 11.3, the search for innovative approaches to tourism in Sardinia is introduced. The following Section 11.4 describes a first experimental application of the multicriteria process, as it could be used to further the involvement of stakeholders and the general public and help to resolve potential conflicts. This same section goes on to present some numerical results, which are discussed and screened by means of sensitivity analysis. In Section 11.5, some remarks are developed on the role criteria and stakeholders play during this multicriteria procedure. In the sixth Section 11.6, a further critical reflection leads to the need to introduce the experimentation of a web-based framework. This is explained in greater detail in the final Section 11.7, where conclusions are derived and new research directions introduced.

11.2 Recent advances in tourism modelling for economic policy and planning

According to approaches stemming from classical economics, tourism per se is believed to bring advantage to the economies of developing countries. The main assumption of these theories is that especially international tourism stimulates a higher level of consumption, thus leading to an overall higher level of disposable income (Krapf, 1961). Inspired by these theories, since the end of the 1960s, many developing countries have introduced international tourism into their economic system (for example, the Caribbean Islands, Mexico, Thailand, Indonesia, the Maldives and Spain). In synthesis, the following characteristics can be associated with the traditional approach to tourist policies: autarchy of tourist entrepreneurship; internationalization; and impacts on local culture that might lead to the loss of cultural identity.

Coccossis and Nijkamp (1996), however, starting from the interaction between tourism and environment, emphasize that, if correctly conceived, tourist activities might contribute to the protection of the natural environment. They point out that classical economic tools like internalization do not seem to solve the dilemma of touristic impacts on local communities. Thus, Briassoulis (1996) claims that mainstream economic analysis is not adequate to support tourism policy decisions, because tourism is not a typical economic sector or activity assumed by this kind of analysis. When tourism is conceptualized as an activity complex, more integrated analytical approaches are required to represent the interrelatedness among the tourism-related economic sectors and the environment. A new paradigm in economics is required for tackling tourism, because today tourism is recognized worldwide as a strategic sector of the economy.

The tourist experience is a multifaceted phenomenon. Research studies (compare Ryan, 1998) investigate the main characteristics of leisure travelling. In the last few years, technological changes, the increase in leisure time, and the specialization of tourism have fostered original issues into the classical patterns of tourist sector planning. The idea is that mono-cultural tourism, based only on the exploitation of the singular beauties of a country, can no longer be considered sustainable. It seems that an innovative model for development should involve diversifying activities and reducing seasonal fluctuation in demand.

Butler (2000) proposes 'complementarity' as a term, similar to sustainable tourism, as it is regarded in an integrative concept. Complementarity is understood as the optimal level of the relationship between tourism and other resource activities. This term implies that tourism and the other activities are not only in relative harmony with each other in the destination region, but in fact add to each other by their mutual presence. The concept will be illustrated in the following section, by using the example of Sardinia.

11.3 A dilemma for Sardinia: Promoting tourism without jeopardizing local resource endowment

The history of tourist activities in Sardinia, one of the main Mediterranean islands, mirrors the shift in the political agenda illustrated in the preceding section. During the 1960s, 1970s and 1980s, the island has been affected by significant financial investments for the construction of its tourist system. The main geographical zones affected still show signs of this relatively big economic push: the north-eastern coastal stretch called the 'Emerald Coast', the south-western coastal boundary, and the south-eastern coastal part. The characteristics of tourist investments and entrepreneurship within this pattern can be summarized as follows: foreign entrepreneurship; foreign management; interest only in the coastal stretch; internationalization of the holiday pattern; introduction of an external model for tourist activities; mono-development of tourist activities exploiting 'sun and sand';

indifference to the suitability of the land for tourist settlements; and little involvement of the local society in decision-making.

As described earlier, in Italy at the end of the 1980s and the entire decade of the 1990s, environmental concerns have induced major conceptual changes in the economics and planning of tourist settlements. In this regard, the trend of tourist integrated policies involves the following strategies: internal entrepreneurship; domestic management; interest for the territory as a whole in its interior and coastal zones; glocalization – linking global networks and local natural and human resources; exploitation of models of tourism stemming from internal human resources; openness to a variety of tourist activities; selection of areas suitable for tourist settlement, as well as high involvement of local communities in decision-making. However, a sort of inertia seems to be affecting tourism in Sardinia and it is still difficult to provide evidence of a diffused movement towards the integration of the tourist subsystem with other economic, social and cultural subsystems.

The main reason for this resistance to the introduction of new strategies is the coexistence of a variety of parties and stakeholders who conceive tourism development from very different perspectives. In this specific case, stakeholders are politicians at the regional and the local level, mayors of the municipalities, and official representatives of private firms or public bodies.

The scenario of the possible models for tourism development suggests the opportunity to reach a compromise solution in which conflicting positions might be successfully reduced and a common framework for tourism development may be provided. In this case, experimentation is developed to assess a system that is useful as a support for decision-making, within the mandatory duties of a generic territorial body responsible for the allocation of funds and investments. Thus, the potential users might be the officials of this hypothetical agency for tourism settlement planning. In their duties, these officials must use tools that take into account a shared model of development for tourist activities.

11.4 Description of the technique

A useful method within the policy perspective introduced in the last section can be found in the family of multicriteria approaches and, in particular, in the set of tools able to tackle complex environmental questions. In regard to consensus building and conflict resolution, multicriteria analysis has a framework suitable for the decomposition of complex values into their simple components. These simple elements can be considered as representations of different concerns and stakes. In this experimentation, the emphasis is laid on the way different points of view can be confronted, various conflicts can be reduced, and intermediate solutions can be sought within a multicriteria framework. In the perspective of an institutional body responsible for territorial governance, multicriteria analysis may become an effective support for decision-making and interactive planning, whenever it stimulates the construction of common knowledge and representation, thus enabling a convergence of the stakeholders towards common views and models for development.

With respect to the operational features of the method selected and studied, many findings indicate that multicriteria tools based on outranking and concordance analysis are usually preferred in environmental integrated evaluations. In this application, the multicriteria method adopted is the qualitative choice method regime (Hinloopen and Nijkamp, 1990), combined with the analytical hierarchy process (AHP) designed by Saaty (1988). The first of these belongs to the broader family of concordance methods developed by Roy (1985). This method has major advantages in comparison to the classical outranking methods (Electre I, II, III), since it allows analysts to process mixed data in an intuitive way and provides the user with the complete final ranking of the alternatives. On the other hand, concordance analysis, allowing for incomparability and incomplete ranking of the alternatives, may lead to fallacies in interpreting the final output.

The method is analyzed in different steps: the specification of the alternatives; the definition of the criteria and the assessment of the weights the involvement of the stakeholders; and the analysis of the results.

11.4.1 The set of alternatives

Since the main objective of the method is to help an institutional body to evaluate the territorial quality with reference to tourism, in this experimentation a set of seven alternatives is considered, as they correspond to relevant municipalities that could host tourist activities. The review of the current state of European funding programmes and of regional special programmes reveals that these domains have many possibilities to receive support. The alternatives consist of the following municipalities located in Southern Sardinia, Italy: Arbus, Pula, Carloforte and Iglesias in the western part of the province of Cagliari, and the Municipality of Cagliari, Muravera and Villasimius in the eastern part of the province of Cagliari. The restriction of the whole stretch of Sardinian coastal municipalities to seven allows for a better understanding of the model. Eventually, this procedure could be extended to the whole set of coastal municipalities. It should be noted that alternatives do not consist of different project options. Rather they refer to different potential characteristics for the seven alternative municipalities, treated as complex values.

11.4.2 Criteria, scores and weights in an adaptive perspective

In a multicriteria analysis, criteria and weights represent the core of the framework, since they mirror what Roy calls the system of individual preferences in the case of a single decision maker (1985). In this situation, the multicriteria approach adopted is used in an open framework to represent a large set of individual preferences belonging to the many stakeholders involved. However, the openness of the framework is not complete. The analyst develops a list of criteria (Table 11.1), discusses the meaning of the criteria, and calculates the weights according to one-to-one interviews with a set of selected stakeholders. The criteria are chosen by confronting different experiences in tourism planning with each other, assuming that a reasonable scheme for the development of local societies can be derived by comparing a number of best practices that have been successful in the Mediterranean area (De Montis, 2001). Therefore, the list of criteria has been set up via an analysis of such case studies.

Criteria have been clustered according to a hierarchy: the general goal, the development of integrated and sustainable tourism, is articulated into seven complex criteria that are themselves decomposed into twenty-six simple criteria. Table 11.1 shows the list of simple criteria with respect to their synthetic name and policy concern. It should be noted that these criteria after having been fixed were not changed during the process. The main consequence of this operational choice is that the scores are fixed. It should be noted that in this chapter the focus is not directed to the scores table.[1] Data have been processed by means of the experimental software 'Samisoft', tested at the Department of Spatial Economics, Free University of Amsterdam.

Following the framework of the analytical hierarchy process (AHP), the weights, meant also as levels of relative importance of each criterion, have been assessed by means of pair-wise comparisons based on the appraisal of a variety of stakeholders. Since the importance of the criteria is recognized to be dependent on subjective feelings and experiences, the set of weights was calculated according to the judgement expressed during a survey of different actors. The survey consisted of separate one-to-one direct meetings with each stakeholder. During the meeting, documents consisting of four parts were presented. In the first part, the *Informative note*, the theme of sustainable tourism was introduced and the whole experimentation explained. In the second part, the *List of criteria*, criteria were presented in their hierarchical structure. In the third, the *Score matrix*, the answers were coded in triangular matrices. In the fourth, the *Questionnaire*, information about the professional activity of the interviewee was gathered.

These materials were presented to 26 stakeholders. The analyst, who aimed to consider a significant number of classes, selected the professionals according to the following criteria. They should be directly or indirectly concerned with tourism policy and planning, should enact the behaviour of a category of professionals, and should have experiences in decision-making in tourism governance. Therefore, the following professionals were selected: six from professional bodies responsible for planning (DE), three officials of environmental and cultural organisations (EN), four freelance professional urban planners (LP), six managers of institutional bodies or of private companies (MG), five public administrators (PA) and two researchers (RE). The list of criteria was presented to each stakeholder, who was asked to compare criteria pair-wise. In this experimentation, substitution of the

---

[1]   For a report on the 7 by 26 matrix, we refer to De Montis (2001).

original tentative criteria list was not allowed: each interviewee expressed his appraisals on the same list.

Table 11.1 General goal, complex and simple criteria

General goal	Complex criteria	Simple criteria		
		Code	Synthetic name	Policy concern
DEVELOPING INTEGRATED SUSTAINABLE TOURISM	DEMOGRAPHIC DEVELOPMENT (C_{DD})	C_{DD1}	Population	Stable settlement
		C_{DD2}	Population growth	Re-equilibrium of population
		C_{DD3}	Human capital	Educating to a highly qualified and varied culture of tourism
	ECONOMIC DEVELOPMENT (C_{ED})	C_{ED1}	Employment	Reducing unemployment starting from critical level
		C_{ED2}	Income per capita	Income distribution
		C_{ED3}	Productivity	Balancing the productivity of the areas
		C_{ED4}	Coherence with EU	Linking operationally tourist projects to EU programmes of financial support
	TOURISM DEMAND (C_{TD})	C_{TD1}	Bed-nights	Balancing tourist bed-nights
		C_{TD2}	Length of stay	Balancing tourist length of stay
		C_{TD3}	Accessibility	Balancing the quality of infrastructure for transportation
		C_{TD4}	Tourist consumption	Balance of the tourist revenues among the areas
	TOURISM SUPPLY (C_{TS})	C_{TS1}	System capacity	Sustain a balanced increase of hotels and residences
		C_{TS2}	Specialized employment	Balancing attitudes to specialization of tourist services
		C_{TS3}	'Second houses'	Recover fiscal benefits
		C_{TS4}	Output in services	Sustaining autonomous development of integrated tourist services
		C_{TS5}	'Tertiary' employment	Encourage tourism within economies of services
	OPERATIVE TOURISM PLANNING (C_{TP})	C_{TP1}	'F' zones	Emphasis of tourist policies within urban and environmental planning
		C_{TP2}	Built 'F' zones	Assign tourist settlements to suitable zones
		C_{TP3}	Carrying capacity	Respecting the equilibrium of local natural resources
	PROTECTION MANAGEMENT (C_{PM})	C_{PM1}	Diversification	Emphasis of tourist policies in non-coastal domains
		C_{PM2}	Park integration	Linking tourist activities to natural parks
		C_{PM3}	Reserve integration	Linking tourist activities to natural reserves
	ENVIRONMENTAL IMPACT (C_{EI})	C_{EI1}	Bathing	Better use of coastlines
		C_{EI2}	Water	Continuous water delivery
		C_{EI3}	Forest	Integrated tourist use of forests
		C_{EI4}	Naturality	Environmental compatibility

11.4.3 The role of the stakeholders

We now discuss the role of the stakeholders during the interviews; the debate about the criteria; the stakeholders' reactions to their structure; the pair-wise comparisons; the level of knowledge of multicriteria evaluation procedures; and the attitude towards Internet-based processes.

Regarding the debate on the criteria, when the stakeholder received the list of criteria, the meanings and concerns associated were introduced to him/her. Often this process of explanation took more than one-third of the global time of the interview: this indicates that the link between meanings and criterion was not so obvious for the stakeholder. The analyst had to face the problem of explaining the pre-designed fixed structure of the list of criteria to the interviewed stakeholders, who often had a completely different concept about the model of integrated tourism and, in the end, about the list of criteria to be proposed. The stakeholders reacted with a wide pattern of behaviour: curiosity, agreement, interest, scepticism, and refusal. Each actor conceived the issue of the integration in tourism in a personal and subjective way. The actors were explicitly asked to agree or disagree with respect to the list of criteria: some agreed, and others disagreed, with the concept embedded in the list. In the case of refusal, the actors could propose a list of criteria which might be different from the one delivered by the analyst. In other cases, the refusal was instigated by the suggestion to change the form of the functions that translate the criteria to numerical figures. In many cases, the interviewees claimed that the structure of criteria, as it was fixed, was too rigid a framework to work with. In all cases, there was a considerable number of suggestions for completing and widening the array of criteria considered.

This is a situation that certainly complicated the process of pair-wise comparison, since the analyst had to go to great lengths to explain the meaning of criteria when he/she asked the interviewees to express the degree of preference. It was very hard for the stakeholder to follow the line of argument embedded in the list of criteria. The main question was due to the difficulty of conceiving a separation of the complex issue into its simple concerns. These difficulties were associated with a general scepticism towards using evaluation frameworks.

As a matter of fact, on the side of knowledge of evaluation schemes, only a few stakeholders had an idea of a multicriteria procedure; most of the interviewees had just heard of the existence of this kind of evaluation model for supporting decisions. They had been taking decisions along unstructured systems of knowledge and on a sort of intuitive rationality.

The analyst, in response to the rigid list of criteria, then presented Internet-based procedures as an alternative. One of the questions was whether the actor agreed to use the Net as a medium for supporting decisional processes. The answers of the interviewees were generally characterized by the tendency to endorse the use of the Internet, even though some scepticism was also expressed. As a whole, the set of stakeholders cannot be defined as a group of 'digiphobes' (Mitchell, 2000), but almost everyone warned about the use of the Net for institutional and mandatory decision-making. The main reason they gave was the still low diffusion of the digital culture, which has been termed the 'digital divide' (Mitchell, 2000).

It should be noted that, during the interviews, the analyst presented the criteria list, discussing the concerns and meanings in dedicated talks consisting of a one-to-one communication between analyst and interviewee. The communication between analyst and interviewee was not affected by any exterior element; and no other stakeholder was involved. Therefore, no debate was allowed among the actors mutually and the analyst. This experience illustrates difficulties of traditional applications of multicriteria decision aid that can arise with regard to incorporating the knowledge of stakeholders and of producing results considered as legitimate.

11.4.4 Discussion of the results

The main result of the procedure is that the analyst elaborated 26 different assessment systems and obtained 26 different and independent sets of weights. Again, this outcome was due to the fact that there was no meeting between the 26 actors. The output of the combination of the weights with the scores yields a 7 by 26 matrix (Tables 11.2 and 11.3), which for each interviewee selected, represents the resulting final rankings of the alternatives.

The set of results is very variable; figures are volatile and change over the categories of professionals involved. In the experimentation, a series of approaches was applied to study and reduce the complexity of this output to a manageable framework. In this chapter only some of these approaches are described: the synthesis of a unique ranking index, the analysis of the aggregate rankings and the frequency analysis of the rankings.

Table 11.2 Final rankings of the alternatives by professional categories[*], part 1

Alternatives	Professional categories codes												
	DE1	DE2	DE3	DE4	DE5	DE6	EN1	EN2	EN3	LP1	LP2	LP3	LP4
Arbus	0.82	0.77	0.65	0.80	0.64	1.00	0.93	0.87	0.75	0.80	0.83	0.47	0.91
Cagliari	0.02	0.35	0.14	0.14	0.39	0.40	0.04	0.19	0.49	0.25	0.23	0.12	0.03
Carloforte	0.50	0.71	0.35	0.35	0.32	0.08	0.32	0.16	0.08	0.27	0.78	0.52	0.44
Iglesias	0.56	0.72	0.93	0.93	0.70	0.29	0.79	0.63	0.52	0.66	0.89	0.81	0.54
Muravera	0.90	0.39	0.89	0.53	0.49	0.42	0.49	0.77	0.66	0.57	0.46	0.61	0.66
Pula	0.42	0.14	0.29	0.25	0.43	0.73	0.31	0.40	0.60	0.80	0.17	0.90	0.60
Villasimius	0.28	0.43	0.25	0.50	0.53	0.58	0.63	0.49	0.33	0.15	0.15	0.06	0.32

Note: [*] For an explanation of the categories, see Section 11.4.2.

Table 11.3 Final rankings of the alternatives by professional categories*, **part 2**

Alternatives	MG1	MG2	MG3	MG4	MG5	MG6	PA1	PA2	PA3	PA4	PA5	RE1	RE2
Arbus	0.43	0.90	0.81	0.79	0.99	0.36	0.52	0.71	0.71	0.66	0.80	0.97	0.41
Cagliari	0.42	0.15	0.03	0.01	0.05	0.20	0.28	0.56	0.46	0.05	0.28	0.01	0.46
Carloforte	0.02	0.36	0.47	0.67	0.73	0.49	0.84	0.29	0.04	0.44	0.22	0.22	0.08
Iglesias	0.67	0.73	0.90	0.45	0.60	0.37	0.36	0.99	0.77	0.53	0.97	0.65	0.14
Muravera	0.81	0.68	0.37	0.54	0.54	0.76	0.79	0.44	0.88	0.65	0.67	0.58	0.81
Pula	0.54	0.15	0.32	0.26	0.40	0.55	0.04	0.29	0.25	0.35	0.40	0.39	0.89
Villasimius	0.61	0.54	0.59	0.78	0.20	0.77	0.67	0.22	0.39	0.83	0.17	0.69	0.70

Note: * For an explanation of the categories, see Section 11.4.2.

11.4.4.1 The synthesis of unique indexes
The ranking symbolizing the aggregated preference of the group of interviewees can be calculated as a vector function of the rankings expressed by each stakeholder. In this case, this function has been adopted as the linear unweighted mean of the final scores expressed by each stakeholder. In such a pattern, the resulting ranking (Table 11.4) consists of the expression of vote, provided that each elector has the same political weight.

Table 11.4 Aggregate final ranking of the alternatives

Alternatives	Aggregate scores
Arbus	0.74
Iglesias	0.66
Muravera	0.63
Villasimius	0.46
Pula	0.42
Carloforte	0.37
Cagliari	0.22

While the group ranks the municipality of Arbus in the first position, Iglesias as second, and Muravera as third, it ranks the main town of the island, Cagliari, in the last position. According to its output, the multicriteria system suggests scenarios where municipalities with underdeveloped social and economic and sometimes also existing tourist systems have to be promoted, especially if they are well-endowed with natural resources.

11.4.4.2 The analysis of the aggregated ordinal rankings
This method has been applied to study the variability of the results, with respect to the different views of the stakeholders. The method consists of two steps. First, the scores corresponding to the stakeholders belonging to the same group have been aggregated, by means of a linear unweighted mean. Second, the resulting aggregate scores have been converted into ordinal terms.

Table 11.5 Variability of the positions in the final average ranking referred to each category of actors

Alternatives	Aggregate scores					
	DE	EN	LP	MG	PA	RE
Arbus	1	1	1	1	3	3
Cagliari	7	6	7	7	6	6
Carloforte	5	7	5	5	5	7
Iglesias	2	2	2	2	2	5
Muravera	3	3	4	3	1	1
Pula	6	5	3	6	7	4
Villasimius	4	4	6	4	4	2

Note: For an explanation of the categories, see Section 11.4.2.

Table 11.5 shows in ordinal terms the different positions occupied in the final ranking by the alternatives, according to each group of stakeholders. The results confirm what Table 11.4 shows: those municipalities that received the highest scores still continue to rank in the highest positions also according to the different groups. Therefore, the municipality of Arbus occupies the first position, according to the evaluation of four groups out of six, and the municipality of Iglesias occupies the second position, according to the evaluation of five groups out of six.

11.4.4.3 The frequency analysis
As in the previous paragraph, the scores, originally expressed in cardinal terms, have been converted into ordinal terms. These figures represent the relative rank of the alternatives for the whole set of interviewees. Thus, it is possible to calculate an absolute frequency matrix, showing the number of times (as a percentage) an alternative has been ranked in a certain position.

In Table 11.6, absolute frequency values refer to the relative number of times interviewees put the alternatives in the different ranks.

Table 11.6 Alternatives versus ranks: Absolute frequencies

Alternatives	Absolute frequencies						
	1	2	3	4	5	6	7
Arbus	42.31	30.77	7.69	3.85	11.54	3.85	0.00
Cagliari	0.00	0.00	3.85	7.69	15.38	30.77	42.31
Carloforte	3.85	3.85	11.54	19.23	26.92	7.69	26.92
Iglesias	26.92	23.08	19.23	11.54	11.54	7.69	0.00
Muravera	11.54	26.92	19.23	34.62	7.69	0.00	0.00
Pula	11.54	3.85	11.54	7.69	26.92	30.77	7.69
Villasimius	7.69	7.69	26.92	15.38	3.85	19.23	19.23

The municipality of Arbus takes the first position according to 42 per cent of the interviewees, and the second position, according to 30 per cent. The municipality of Iglesias has been ranked in first position according to 27 per cent, and in second position according to 23 per cent. The municipality of Muravera is placed in the first position according to 11 per cent of the interviewees, and in the second position according to 27 per cent.

It is not surprising that these results confirm the pattern stemming from the ranking of the mean of the scores, as displayed in Table 11.4. This evidence again shows that municipalities with a rich natural endowment are placed in the highest positions by quite a large share of the interviewees.

11.5 A critical reflection on the process

The variability of the set of resulting scores depends on the semi-openness of the entire multicriteria procedure. In other words, the rankings are variable, because the weights have been calculated with reference to the judgements expressed by the different stakeholders. Yet, by fixing the list of criteria, the analyst may have induced a sort of 'environmental bias' in the trend of the final rankings: the municipalities with a greater endowment of environmental resources rank in the highest positions. The selection procedure for the criteria might also have led to a generally high sensitivity of the results, since each interviewee reacted only according to his/her own interpretation of the criteria. This is confirmed by the mode of communication between analyst and interviewee. In a direct one-to-one dialogue, there are not many chances to enrich the discussion with contributions coming from other fields and to develop a convergence regarding a common point of view about criteria and weights. In terms different from multicriteria technicalities, it could be said that the construction of a shared vision on the model of tourism development in these conditions may perhaps be more a hope than a feasible option.

Moreover, the analyst again, with reference to the 'main' actors playing in the scene of tourism development, has decided on the set of stakeholders. They can be considered as representative of six categories of professionals, but there is no statistical evidence of this relationship. Indeed, even when a fair representation may be assumed, the number of categories considered should not be limited to six. On the other hand, the representation should be extended to the many other communities, which could not have the same 'voice', but should still be encouraged to express their opinion on the future scenarios of development based on tourism. In this way, a variety of concerns might arise that may disturb the rigid list of criteria and enrich the originally steady portrait. Within this option, the analyst would become more a communicative master of the debate, while each actor would exchange opinions with the other stakeholders.

These points reveal the need to individuate places, sites, and halls, where the debate can be addressed to building common points of view regarding concerns, objectives, and meanings of tourism-based development policies. To this end, an interactive participatory process based on the Internet has been designed by constructing a virtual decision arena to facilitate the involvement of more stakeholders in the decision-making process. In further studies, the Internet-based procedures will experiment with large and different samples of users.

11.6 Virtual environments for communicative decision making

The experimentation that will be now described has been designed to allow for new digital forms of collaborative planning. In this environment, Internet distributed computing is applied to negotiating processes where decision making is coupled with group learning and distributed computing.

This application can be considered one possible interpretation of the concept of cyber-multicriteria decision-making. The procedure consists of processes that are made accessible via Internet hyperlinks and is illustrated by means of a demo-simulation available on request.

The system consists of a virtual decision-making arena, which allows group leaders to manage a variety of activities, such as group learning and interaction. The info-design of a complex Internet interface provides an interactive spatial decision support system, through which each member of the group has the opportunity to enter a communicative arena and carry out all the activities and debates described in Section 11.5 above. Within this framework, the stakeholder can became familiar with the sensitivity of the procedure, propose modifications by indicating new criteria and even a different approach to evaluation.

11.6.1 The digital lab 'Evaluating Tourism'

The application introduced above has been developed on the basis of the research advances in multicriteria methodology.

The above-mentioned concerns about communication inside the process have led to the design of an Internet application, the project called 'Evaluating Tourism', which we believe will allow for an 'on-line evaluation' process, based on a remote access debate among the stakeholders. The complexity of the management of the process, where many actors have the opportunity to communicate, had led to the integration of multicriteria analysis with information and on-line communication technology inside a single environment.

The experimental Internet site has two main areas of service: a public and a private domain. This choice meets the need for allowing access to the community as a whole and for opening virtual laboratories dedicated to the activities of the decision-making group. The Home Page presents the available fields for action: an area for geographic information retrieval; a public forum; and a private consultant's area. The information about the nature of the site, or the general 'message', is formulated so that a normal user can understand the main concepts and scope of the procedure, without having to understand technicalities.

The public domain of the site presents a discussion forum for users, where simple geographic information can be retrieved, accessible through the area 'GIS on-line' (Figure 11.1). In this area, geographic maps can be examined and the user can navigate through them with zoom and pan commands. These features are designed to allow for easy consultation by non-expert users.

The domain of the public Forum is regulated by characteristics that are common to Internet discussion groups: free access allows the users to choose the fields being debated and to express their opinions. The virtual agora takes shape spontaneously, and collective intelligence emerges from a variety of judgements stimulated by society as a whole. In this Forum, it is possible to gather useful information about citizens' images of the development of local tourism. The use of this system could help to suggest how a further debate mechanism could be elaborated to stimulate the exchange of opinions. Following this strategy some topics could be proposed, while citizens could be encouraged to debate other topics in a multiplicity of 'virtual living rooms'.

The private domain of this site is only directly accessible through the Home page for those decision-makers who have been given an account by the process administrator. The system of virtual activities is accessible through the list of links in the welcome page (Figure 11.2). The link 'Il Progetto' (The Project) allows access to information about the strategic objectives of the evaluation process for the development of sustainable tourism in southern Sardinia. Many activities can be coordinated in this electronic environment: knowledge construction, evaluation procedures, and schedules of meetings.

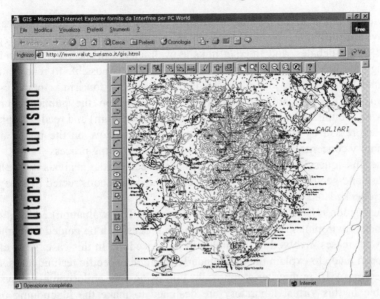

Figure 11.1 An example of 'zooming' in to a geographic area in Southern Sardinia through the GIS on-line function

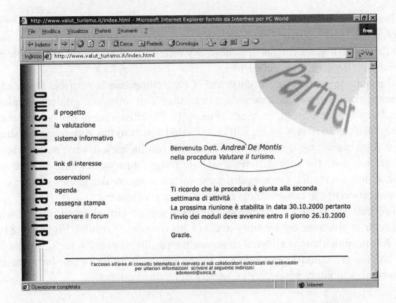

Figure 11.2 The welcome page of the private domain: On the left of the screen a list of virtual activities is shown

Knowledge construction is supported by a series of links. Useful sources are displayed in different ways: information from public debates, newspaper reports, public forum debates, and Internet links as well as an on-line GIS (Figure 11.3). Through the link 'rassegna stampa' (Press review), it is possible to read materials contained in electronic journals related to the issue of tourism. In the private domain, each participant can retrieve information from the public Forum by following the link 'osservare il forum' (Watching the forum) and read the opinions being expressed by the citizens. This is a sort of window on the public arena: community concerns enrich elements of the decision-making process.

The link 'query' allows one to perform advanced spatial analysis by means of database and geographic filters (Figure 11.4). The data is constructed in a way that can be checked and verified.

The section related to the link 'la valutazione' (The evaluation) consists of an electronic environment where the actors in the process can be guided throughout a variety of phases involved in the multicriteria procedure. In this case, great efforts have been made to explain the logic underlying highly specific techniques, such as the Regime-AHP multicriteria approach. Following this philosophy, the systems available in this virtual laboratory are designed to make the functioning of the techniques easier to understand. The continuous interactions between the actors and the system helps in the process of harmonizing the conceptual significance of all of the evaluation steps, such as building the criteria system and debating the results.

The link 'I criteri' (The criteria) leads to a virtual lab where evaluation criteria are presented and compared with the system of objectives and the values associated with them. Each actor can see the hierarchy of criteria at all times and can propose corrections (Figures 11.5 and 11.6).

The links 'I pesi' (The weights) and 'Come compilare la matrice' (How to fill in the matrix') explain the calculation procedure and introduce guided completion procedures for the criteria matrices (Figure 11.7). This process evolves by means of the electronic delivery of digitally completed forms to the system administrator. The system elaborates these data and sends the results back to each actor, who can thus realize how the multicriteria evaluation algorithm works and, if necessary, propose improvements. Once again, these interactions are designed to increase the harmonization of the group for the areas under consideration.

The link 'I risultati' (The results) opens a sort of virtual scrutiny hall, where each actor is given the opportunity to check his own set of results (Figure 11.8) and only the administrator is allowed to screen the results of every actor (Figure 11.9). The last feature could be extended to all the actors, in order to test how sensitive the system is to interference from a variety of results systems.

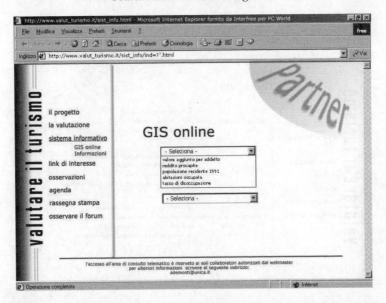

Figure 11.3 A simulation of a spatial query, from the input of the data requested

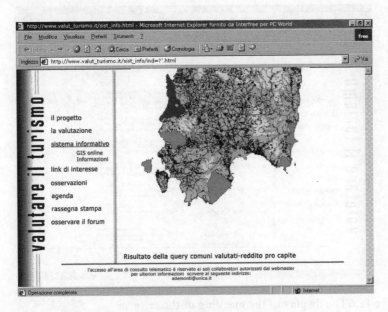

Figure 11.4 Display of the results about the evaluated areas

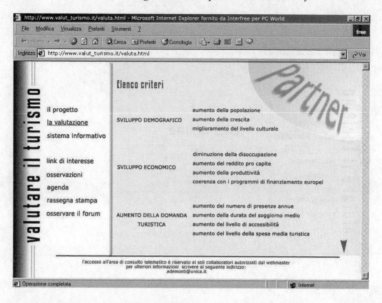

Figure 11.5 The display of criteria hierarchy

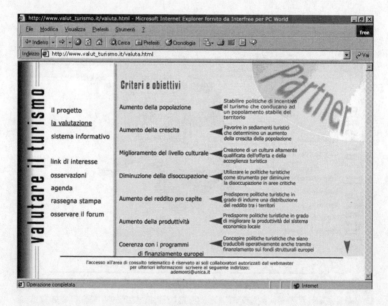

Figure 11.6 The display of the meaning of the criteria

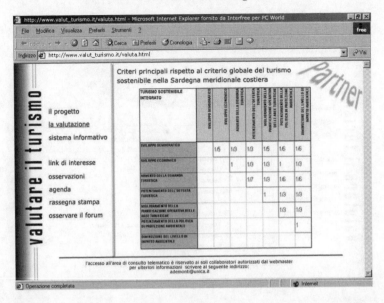

Figure 11.7 The activity of filling in the matrix is guided by the instructions displayed

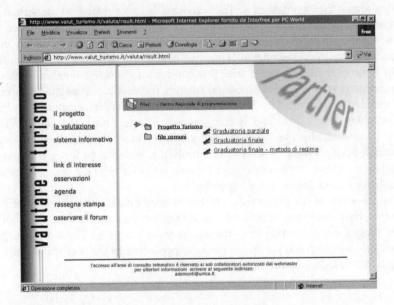

Figure 11.8 The display of results for each stakeholder

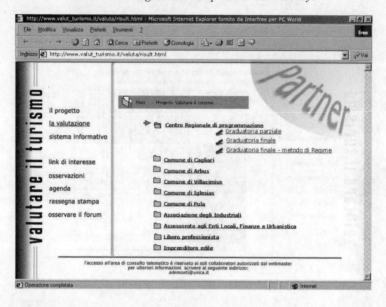

Figure 11.9 The display for the evaluation administrator

The virtual spaces described do not simply duplicate the traditional decision-making arena: they strengthen it. The activities that take 'place' in the virtual labs dedicated to evaluation contribute greatly to stimulating participation, since they allow each actor to be intensely aware of his/her cognitive participation in the process and ability to visualize it. Thus two domains of activities can be identified in the same evaluation process. The first consists of a discrete number of face-to-face meetings with a 15-day schedule, where a 'contact' confrontation evolves in the traditional manner. The second consists of continuous meetings in digital space, where interactions and learning evolve in a tele-mediated pattern. These fields can actually be considered complementary aspects of a single communicative planning process. Virtualization is essential to foster learning and negotiation. These continuous background activities prepare an environment suitable for shared decision-making paths.

At this stage of the project, the aforementioned public and private domains of activities have been just conceived, since they are projected on the basis of the many suggestions stemming from the interviewees reactions. The authors hope to extend this approach and set up a series of experiments on the practical usability of the Internet-based multicriteria framework.

11.7 Concluding remarks and open questions

While some general comments have already addressed the results of the multicriteria procedure, this section is focussed on its possible web-based solutions. This tool represents an example of a system-user interface, designed using techniques for the construction of multimedia communication and virtual reality. In this view, the system can be described as a communication management system that is available either by public remote access (Internet pattern) or by password protected remote access (Intranet pattern)[2].

'Evaluating Tourism' consists of a co-ordinated network and information system, in so far as it supports the management of interactions and communication between the analyst and other actors in the evaluation process. In terms of communication, it is a useful tool, not only because it leads to a reliable set of results but also because it is able to foster a lively debate among the stakeholders. The deliberate reconstruction of the local development model is the basis for a shared process of decision making. These results reveal new 'agoras' for planning, where the 'cyber process' has the opportunity to evolve. The experimental virtual decisional lab 'Evaluating Tourism' is designed to allow for faster interactions and more personalized evaluation paths in a real decisional hall.

The main utility of the internet-based procedure is that it is able to reduce the level of conflict by stimulating the debate among the actors. The site allows direct use of the system of multicriteria analysis leading to a reduction of the subjectivity of the results, since the actors may converge on a common way of conceiving concerns and objectives.

In addition to the aforementioned results, there still seem to be some questions that remain open for future research. The first concerns the extent to which a society and its planners are capable of adapting to innovations in communication technology and 'digital planning'. The digital age seems to produce not only a change in the mode of communicating, but also a mutation in a particular community's patterns of knowledge and existence. Indiscriminate expansion of the availability of digital planning aids could create an inflationary perspective for cyber planners in terms of planning practice. Planners might risk becoming the victims of programmers of electronic sequences that are not necessarily related to the real organization of their geographical spaces. This is the main reason for concern about the widespread diffusion of video-game patterns inside many available software packages for environmental and urban simulation. The term 'Nintendo decision-making', coined by Carver (1999) with reference to the practice described above, depicts the situation of contemporary planners, who may be too absorbed by the innovation of digital simulation to focus on the real consequences of actual changes in real geographical places.

---

[2]   A demo-simulation of the experimental site 'Evaluating Tourism' is available on request. The readers may refer to the authors of this chapter.

Even though these scenarios raise disquieting questions about what tele-presence is and what potential human sensory capacities are, there nevertheless seems to be a need to develop further studies on the possible ways of improving planning procedures through the use of digital technologies.

In particular, there seems to be widespread optimism about the ability of humans to interpret ideas coming from virtual spaces. In this case study, the development of the system of virtual laboratories' potential, which supported the evaluation process, may allow the meaning of a specialized technique such as multicriteria analysis to be communicated to the group.

In further studies, the practical use of the Internet-based features will provide the authors with important information about the efficiency in the transmission of all the 'messages' connected to a communicative multicriteria planning support system.

The second question, which is connected to the first, concerns the new role of decision-making centres in light of the possibilities offered by digital tele-presence. The issue can be divided into two parts: planning theory and planning processes.

From the point of view of planning theory, the advent of 'cyberspace' and of the opportunities linked to the concept of simultaneous presence in many places might lead to a lessening of the sense and perception of territory (Mitchell, 1995). Classical categories of Euclidean space have to be re-interpreted, since electronic spaces have different topologies. In sociological terms, one of the main results can be seen as a change in the sense of belonging to the same 'territory' as a sign of cultural identity. The Net, interpreted as a system able to connect a theoretically infinite number of places, might suggest the idea of loss of territory, known as 'de-territorialization'. Time-space substitutes for geographical space: the real-time aspect of the Internet can remove the barriers of physical distance, with the eventual possible disappearance of the concept of geographical place, and thus of borders that have hitherto been the foundation of a community's social identity.

The possibility of real-time presence can encourage a radical change in the structure of planning processes. Planners are currently faced with new digital tools and instruments that seem to mean that temporal and spatial patterns of work must change. At the same time, it seems that the job description of the contemporary planner will be characterized by requiring expertise in a variety of innovative fields, such as geo-informatics, remote sensing, pattern recognition, objects programming, and process programming.

Furthermore, social landmarks seem to be in the process of mutation as far as the construction of the master plan is concerned. According to recent radical perspectives, if a collective intelligence established itself within the framework of the master plan, each individual citizen would be able to elaborate and propose solutions for the use of spaces. The application of direct democracy mechanisms to planning practice would mean the re-organization of the functions and duties not only of decision-makers and communities, but also of analysts and planners.

The emerging dissolution of the territory and of the planning process may be a problem for the future of planning. However, the evidence found by this chapter supports a more optimistic position. A planning decision cannot be envisaged without an associated, geographical location. No virtualizations of the territory can substitute for perceptions or real experiences of it.

The approach advocated in this chapter seems to provide a framework of valid suggestions, which could stimulate further research on the innovative role planners could play in interpreting current changes in information technology.

References

Associação Portuguesa para o Desenvolvimento Regional (APDR) (2000), 'Tourism Sustainability and Territorial Organisation', *XII Summer Institute of the European Regional Science Association*, Grafica de Coimbra Lta, Coimbra, Portugal.

Briassoulis, H. (1996), 'The Environmental Internalities of Tourism: Theoretical Analysis and Policy Implication', in H. Coccossis and P. Nijkamp (eds), *Sustainable Tourism Development*, Ashgate Publishing Company, Aldershot, UK.

Butler, R.W. (2000), 'Tourism, Natural Resources and Remote Areas', in Associação Portuguesa para o Desenvolvimento Regional (APDR), 'Tourism Sustainability and Territorial Organisation', *XII Summer Institute of the European Regional Science Association*, Grafica de Coimbra Lta, Coimbra, Portugal, pp. 47–60.

Carver, S. (1999), 'Developing Web-based GIS/MCE: Improving Access to Data and Spatial Decision Support Tools', in J.C. Thill (ed.), *Spatial Multicriteria Decision Making and Analysis: A Geographic Information Sciences Approach*, Aldershot, UK, pp. 49–75.

Coccossis, H. and P. Nijkamp (eds) (1996), *Sustainable Tourism Development*, Ashgate Publishing Company, Aldershot, UK.

De Montis, A. (2001), 'Valutazione e pianificazione per la gestione dei processi insediativi. Dall'approccio "razionale" all'approccio "comunicativo"', Ph.D. Thesis. Dipartimento di Architettura e urbanistica per l'Ingegneria, Università degli Studi di Roma, La Sapienza, (unpublished).

Hinloopen, E. and P. Nijkamp (1990), 'Qualitative Multiple Criteria Choice Analysis: The Dominant Regime Method', *Quality and Quantity*, 24: 37–56.

Krapf, K. (1961), 'Les pays en voie de développement face au tourisme: introdution méthodologique', *Revue de Tourisme*, 16(3): 82–89.

Mitchell, W.J. (1995), *City of Bits, Space, Time and the Infobahn*, MIT University Press, Cambridge, USA.

Mitchell, W.J. (2000), *E-topia: Urban Life, Jim – But Not as We Know It*, MIT University Press, Cambridge, USA.

Pearce, D.G. (1989), *Tourism Development*, Longman, London.

Roy, B. (1985), *Méthodologie Multicritère d'Aide à la Décision*, Economica, Paris.

Ryan, C. (ed.) (1998), *The Tourist Experience*, Redwood Books, Trowbridge, Wiltshire, UK.

Saaty, T.L. (1988), *Decision Making for Leaders: The Analytical Hierarchy Process for Decisions in a Complex World*, RWS Publications, Pittsburgh.

Wall, G. (1997), 'Sustainable Tourism – Unsustainable Development', in S. Wahab and J.J. Pigram (eds), *Tourism Development and Growth: The Challenge of Sustainability*, Routledge, London, pp. 36–52.

Chapter 12

A Methodology for Eliciting Public Preferences for Managing Cultural Heritage Sites: An Application to the Temples of Paestum

Patrizia Riganti, Annamaria Nese and Ugo Columbino

12.1 Introduction

The museum sector has been the object of increasing interest in the last ten years, as shown in several publications (Jackson, 1988; Frey and Pommerehne, 1989; Feldstein, 1991; Frey, 1994). Many research studies (i.e. Silbeberg, 1995; Verbeke and van Rekom, 1996; Harrison, 1997; Johnson and Thomas, 1998) have focussed on museums' services, acknowledging the importance of the aspects related to public fruition[1] over those mainly targeted to solely fulfil the exhibition purpose. Within this framework, and in tune with the understanding of the social role played by art, museum management issues have been increasingly linked to market dynamics, showing the need to understand public preferences. In fact, financial investments in the museum sector can be better justified when related to improvements in public fruition and in the understanding of the art piece. The contingent valuation method is a survey-based valuation technique that, because of its nature, has the potential to be very participative. People can express their preferences for non-market commodities stating their willingness to pay (WTP) for changes in the provision of the good. In this way, the latent demand curve for the good concerned can be traced. Recent literature shows several examples of applications of the contingent valuation method to cultural goods. A more restricted number of studies focus on the use value of museums. Ashworth and Johnson (1996) analyze the monetary value that individuals attach to the museum visit; Scarpa et al. (1998) elicit the value of access to the Contemporary Art Museum of the Rivoli Castle near Turin, Beltran and Rojas (1996) estimate WTP

---

[1]   The term 'fruition' refers to both physical access to, and appreciation of, the cultural good concerned.

for the fruition and conservation of some archaeological areas in Mexico; and Mazzanti (2001) elicits the WTP for the conservation of the Borghese Gallery Museum in Rome and for the introduction of some new services, e.g. increase in opening hours, multimedia service and non-permanent exhibitions.

The research reported in this chapter aims to contribute to the current literature debate on the method, using the conjoint analysis format to elicit the level of desirability of different management policies for the services in support of the Temple of Paestum's archaeological area and its museum. In particular, we analyzed alternative policies focussing on different ways of experiencing the good. We considered three different policy packages, the first mainly concerned with improvements in the fruition aspects; the second mainly targeted at leisure time; and the third aimed to enhance educational purposes.

A sample of 732 respondents was gathered at the site in order to elicit individual users' preferences for different management options of the site. Each respondent was presented with three different scenarios (A, B and C), each differing from the others in terms of the kind of museum service provided, and the entry fee. Each scenario constituted an alternative management option, corresponding to the following broad categories: a) mainly fruition, meaning accessibility to the different parts of the site and its museum as well as improvement in the understanding of the cultural good; b) entertainment; c) education.

The study was funded by the Regione Campania, the local authority, and was part of a research project devoted to the study of economic models for the management of cultural heritage goods. Some of the most desired attributes considered in the analysis are now being implemented.

The chapter is structured as follows. Section 12.2 describes the main problems in managing cultural sites and the potential of methods such as conjoint analysis in eliciting public preferences; Section 12.3 describes the questionnaire, the survey implementation and the main statistics of the selected sample; in Section 12.4, the theoretical and the econometric model are discussed and the results presented; and, finally, Section 12.5 provides our conclusions.

12.2 Management of cultural sites and public preferences elicitation

12.2.1 Managing cultural heritage in the perspective of sustainable development

Cultural sites represent an increasingly important economic resource for the development of a region. Cultural tourism is now spreading in many European regions, thanks also to the new air travel opportunities given by low-cost airlines. The development of cities needs to account for the necessity for the appropriate management of cultural goods to be sustainable in economic, cultural and social terms. The role of valuation techniques becomes prominent in this context. How to assess management strategies for cultural goods conservation is a matter of the research and enhancement of current valuation methods.

This chapter discusses one of the possible approaches to cultural heritage management, based on the elicitation of public preferences concerning the economic values of intangible goods, which are usually considered unpriced. The methodology used here refers to the economics of outdoor recreation and emphasizes the use of contingent valuation, one of the economic valuation techniques developed during the twentieth century by environmental economists.

Managing cultural heritage sites implies finding optimal ways to combine the conservation aspect with the need to generate income from the site. In turn, this requires the use of valuation methods to assess more preferable options. Since the Athens Charter (1931), the role of historic building conservation has been highlighted at international level. There then followed a number of other international documents, such as the Charter of Venice in 1964 and the Granada Convention in 1985, which stress both the relevance of the attached economic values and the importance for city development of entire cultural sites. Other international agreements have since then highlighted the need for the integrated conservation of cultural heritage both in terms of buildings and of sites (Declaration of Amsterdam 1975; Washington Charter 1987).

The Venice Charter in 1964, for the first time, saw cultural heritage sites as economic goods, and therefore a resource, and an asset. More recently, participants at the meeting of UNESCO and the World Bank in Beijing in July 2000 with experts from all over the world, emphasized the relevance of regulations as a prerequisite for the protection of cultural heritage that needs to involve both decision-makers and local communities. On this occasion, the debate confirmed that the preservation of cultural heritage has long been perceived as a 'public expenditure therefore excluded from cost/benefit analysis' (Luxen, 2000). There is the need to develop a new attitude, where preservation and restoration works may be perceived as real investments. The acknowledgment of the economic values attached to cultural goods is of strategic importance in order to change the negative attitude at policy level.

12.2.2 The potential of conjoint analysis

In the last few decades, environmental economists have developed non-market valuation techniques to elicit public preferences in the form of economic values attached by the relevant population to policy alternatives. These techniques aim to compute the monetary benefits of environmental policies, important when one wants to compare different categories of benefits, or when one wants to compare the benefits of a policy with its costs.

When one wishes to place a monetary value on the unpriced features of a cultural site using stated preference techniques, two approaches are possible: contingent valuation (Mitchell and Carson, 1989), and conjoint choice studies (Hanley et al., 1998).

In a contingent valuation survey, people are asked directly to report their willingness to pay (WTP) to obtain a specified commodity, such as the way a conservation site is managed. The proposed change is generally hypothetical, and no actual transaction takes place. Contingent valuation has been traditionally used to place a monetary value on environmental goods. More recently, programmes for the preservation and restoration of specific sites or *buildings* with historical and cultural significance, such as churches, museums, theatres, and marble monuments, have been valued using this technique. A survey of some studies can be found in Navrud and Ready (2001). A more extensive review of the main studies is in Noonan (2002).

Conjoint choice analysis can be considered as a more recent development of the contingent valuation approach, and seems even more suitable for management purposes. In a typical conjoint choice experiment study, respondents are asked to choose between two or more commodities (or 'policy packages'), each of which is defined by a set of attributes, one of which is usually the cost to the respondent. Attributes are varied across 'packages', and the packages are usually matched in such a way that respondents must trade off attributes to make their choice. Conjoint choice analysis, therefore, seems potentially the best valuation technique when the aim of the valuation exercise is the assessment of changes in policies or programmes. Alberini et al. (2003) test the technique's suitability to assess urban regeneration projects, aiming to elicit local residents' preferences for alternative options. Conjoint analysis has so far shown to be flexible enough to accommodate different policy purposes. To this extent, the major challenge posed to researchers is represented by the hypothetical scenario's definition, in terms of the appropriate attributes, and their levels, identifying the choice. This chapter provides an example of how this technique might provide monetary estimates to support decision-making for site management.

12.3 Eliciting preferences for a world heritage site: The Temple of Paestum

12.3.1 The archaeological site

This study focusses on the archaeological area of Paestum and its museum. The Temples of Paestum: namely, the Basilica, the Temple of Poseidon, and the Temple of Ceres, are among the most impressive examples of archaic Doric architecture outside Greece. They were built between 530 and 460 BC as part of the city of Paestum, one of the most important Greek colonies in Magna Grecia. They were inscribed in the UNESCO World Heritage List in 1998, within the Cilento and Vallo di Diano National Park, together with the archaeological sites of Velia and the Certosa of Padula. They are among the most important archaeological remains in Italy and are visited by many tourists. A museum situated next to the archaeological remains contains many Roman and Greek works of art.

The conjoint analysis study presented here was a response to the local political agenda of developing new management policies for the conservation and valorization of this outstanding site. An increase in the level of fruition and understanding of the Temples of Paestum, and the role played by them in the whole region, might encourage tourists to visit other nearby cultural sites as well. A sensible increase in tourist numbers was therefore welcomed, if this also meant redirecting tourists to other nearby archaeological areas and transforming the day trippers into resident tourists for an overnight stay. This would bring economic benefits to the development of the entire area. As discussed above, conjoint analysis appeared to be the most flexible and adequate valuation technique for eliciting preferences. At the time of the questionnaire's development and the first survey's implementation, there were no similar studies available in the literature. Contingent valuation had been used for a number of cultural goods (Noonan, 2001; Navrud and Ready, 2002), but the museum sector had been almost ignored. During the course of present research, other studies were conducted on similar topics and more recently published (Santagata and Signorello 2000; Mazzanti, 2001).

12.3.2 The questionnaire and the survey implementation

A crucial aspect of any conjoint analysis is the development of an appropriate questionnaire. For our study, we followed the usual steps envisaged by the literature. First, two focus groups were held in June 1999 which aimed to understand the kind of services which were particularly preferred by the local population. Then two pre-tests were conducted, one at the end of June 1999 and the other in mid-July 1999. The final version of the survey was implemented in August 2002.

The pre-tests and the final survey were all carried out on site. The first pre-test consisted of 50 interviews collected by five interviewers. The second pre-test consisted of 245 interviews gathered on site by the same five interviewers. Major changes were made in the questionnaire wording and structure between the first and the second pre-test, while only minor changes were envisaged after the data analysis of the second pre-test. The final survey was carried out in August 2002 on site by seven interviewers who gathered 732 interviews.

The final questionnaire consisted of four major sections to be administered to the respondent, plus two sections to be filled in by the interviewer. The first section included questions eliciting the respondent's attitude with respect to the category of goods being valued: namely, cultural goods. The second one presented the description of the good, the archaeological area of Paestum, and some questions aimed to elicit the level of the respondents' knowledge of the good. The site description was as, is usual, reinforced by photographic images and maps collated in a brochure prepared in collaboration with the Sovrintendenza, the local agency in charge of the site's conservation. The third section consisted of the valuation question, in this case in conjoint choice format. The fourth section included questions eliciting the major socio-economic characteristics of the respondents (age, sex, income, level of education, etc.). The two final sections were filled in by the interviewer and included comments on the respondent's attitude throughout the interview, plus other relevant information.

Great care was devoted to developing the valuation question part that is obviously the crucial one to elicit monetary expressions of respondents' preferences. Each alternative was given by the combination of different levels of the attributes defining the scenario. In our choice experiment, we had nine attributes plus the cost of the 'package'. We randomly derived a combination of alternatives, to be shown in pairs to the respondent, taking care to eliminate the dominated ones and checking for the appropriateness of the level of the attribute cost (in order to avoid the possibility that packages with more expensive services might be 'sold' at cheaper prices). We generated 24 cards each showing three options, one of which corresponded to the minimum number of services representing the site conservation option (Scenario A). Each respondent was required to express his/her preference among the three options (Scenarios A, B and C), where Scenario A did not assume any extra cost compared with the current ticket price. The choice experiment was repeated 4 times per each individual. The card order was regularly rotated in the administration of the sample in order to avoid ordering bias.

The attributes composing each of the scenarios fell into three main categories: a) fruition services, that is, improving the accessibility and understanding of the site; b) leisure services; c) educational services. Table 12.1 shows an example of a preference card.

Among the services targeted at improving accessibility, we have: an increase in opening time (from 9am to 10pm, instead of sunset); audio guides with a recorded description of the museum and the archaeological site; and hourly guided tours. The services targeted at educational purposes are: a children's lab; a multimedia reconstruction of the archaeological remains; and an IT documentation centre on the other archaeological sites of interest in the surrounding region. The leisure services include: a café within the archaeological remains; the organization of weekly concerts/performances; and the provision of non-permanent exhibitions. The cost to the respondent varies between €6.20 and €12.91.

12.3.3 The data

Table 12.2 presents the socio economic statistics of the sample of respondents used in the econometric analysis. The sample consists of 552 observations, after the elimination from the samples of the observations with missing information on one or more of the crucial variables.

The information reported in Table 12.2 indicates that the sample is mainly composed of people resident outside the Campania region (76 per cent), with a good level of education (54.12 per cent of the individuals had completed secondary school, and 39.61 per cent were graduates or more highly-educated). The majority of individuals reported a household gross income of between €20,000 and €30,000, while 42.82 per cent of the respondents indicated an income higher than €40,000. Most individuals were aged between 24 and 48, while the percentage of people between 18 and 23 years of ages was quite low (about 7 per cent).

Table 12.1 An example of a preference card

Attributes	Scenario A	Scenario B	Scenario C
Opening hours	From 9am till one hour before sunset	From 9am to 10pm	From 9am till one hour before sunset
Audio-guides for the archaeological remains and the Museum (Not included in the entrance fee)	NO	YES	YES
Experts guided tours (Not included in the entrance fee)	NO	NO	YES
Café with view on archaeological remains (Purchase not included in the entrance fee)	NO	YES	NO
Thematic non-permanent exhibition (Access not included in the entrance fee)	NO	YES	YES
Weekly cultural Events (classical/pop music concerts and theatrical performances) from June to September (Access not included in the entrance fee)	NO	YES	NO
Children's Lab (Access not included in the entrance fee)	NO	NO	NO
Audiovisual projections along the museum and site itinerary (Use included in the entrance fee)	NO	YES	YES
IT documentation centre (Use included in the entrance fee)	NO	NO	NO
Price	€6.20	€7.75	€12.91

As described in Section 12.2, individuals in the sample were required to express their preference among three scenarios, where Scenario A did not assume any extra cost to the current ticket price, which corresponded to the minimum number of services needed for the site conservation. As shown in the last row of Table 12.2, only a relatively small portion of respondents people (8.69 per cent) selected Scenario A, involving no extra cost.

Table 12.2 Descriptive statistics (N=552)

Variable		%
SEX	Female	46.67
	Male	53.33
RESIDENCE	Paestum	1.97
	Campania	22.05
AGE	18-23	7.53
	24-28	14.34
	29-33	13.80
	34-38	18.64
	39-43	15.59
	44-48	13.98
	49-53	6.63
	54-58	5.02
	59-63	1.97
	>63	2.51
EDUCATION	Compulsory level or less	6.27
	High school	54.12
	University	39.61
HOUSEHOLD INCOME	<10	4.12
	10-20	18.64
	20-30	34.40
	30-40	21.86
	40-60	13.08
	>=60	7.88
RESPONDENTS WHO SELECTED SCENARIO A		8.79

12.4 The model and the results

12.4.1 The theoretical and econometric model

In this section we discuss the econometric model used to analyze our data set. We assume that the monetary value V(*s*) of a certain scenario *s* can be obtained by the following expression:

$$V(s) = \beta_1 X_1(s) + \beta_2 X_2(s) + ... + \beta_9 X_9(s) - \delta PRICE(s) \tag{1}$$

where the $X(s)$ indicate the first nine characteristics of the scenario *s* (e.g. $X1(s)$ = hours in scenario *s*; $X4(s)$ = 1 or 0 whether the café is present or not in scenario *s*); and PRICE(*s*) is the ticket price associated with scenario *s*.

The estimated coefficients β divided by δ give the monetary value, willingness to pay (WTP), attached by the respondent to the service identified by β. For instance, β_1/δ is the WTP for an increase in the opening hours, while β_4/δ is the WTP for the presence of a café within the archaeological area. We assume that the respondent is capable of accounting for his/her budget constraint and of making the appropriate trade-offs with the other attributes of the choice set, when choosing the preferred option.

If we consider the increase in revenue related to the ticket price increase and the increased number of tourists, we can determine to what extent a certain policy can cover maintenance costs. One can show that the following result holds.

The estimated variation in the visitor numbers when one moves from scenario '0' to scenario 'S', characterized by a certain increase in the number of services, can be calculated as follows:

$$100(\exp\left\{\sum \beta_i(X_i(S) - X_i(0)) + \delta(PRICE(S) - PRICE(0))\right\} - 1) \qquad (2)$$

Such analysis can prove very useful, given the current attitude of museum management to introduce new information technologies (computer, video, audio guide, etc.) in support of fruition, often at very high costs. These costs might appear more reasonable if one could prove that they are at least partially met by the generated cash flow.

12.4.2 The results

Table 12.3 presents the coefficient estimates, standard deviations, and 't' values obtained using our data set.

The results show that respondents attach a significant and positive value to all characteristics presented in the choice set, except the café, which seems to be perceived negatively. The most preferred services are guided tours, an increase in opening hours and a children's lab. Less interest is shown for performances, concerts, and non-permanent exhibitions. Among the educational services, the smaller WTP is attached to the documentation centre on the archaeological sites present in the region. Using expression (2), one can calculate the increase in visitor numbers caused by the introduction of a specific service.

Previous studies have elicited individual preferences for museums, though with different approaches (e.g. Ashworth and Johnson, 1996; Beltran and Rojas, 1996; Mazzanti, 2001; Santagata and Signorello, 2000), and have found that the interest in cultural goods is linked to individual characteristics such as income, education, sex, and age. Table 12.4 presents the WTP for each service as a function of household income, and, as expected, the marginal utility is decreasing.

Table 12.3 WTP for the different services (552 observations)

Variables	Coeff.	Std. Dev.	t-stat	β/δ*
HOURS	0.6580	0.0747	8.812	5.22
AUDIO	0.4594	0.0634	7.240	3.64
TOURS	0.8018	0.0675	11.881	6.36
BAR	-0.1937	0.0734	-2.639	-1.54
EXHIBIT	0.2936	0.0733	4.007	2.33
EVENT	0.4561	0.0764	5.972	3.62
LAB	0.7025	0.0805	8.723	5.57
AUDIOV	0.4880	0.0738	6.608	3.87
DOCUM	0.3979	0.0622	6.401	3.16
PRICE	-0.1260	0.0069	-9.444	

* marginal WTP in euros for the museum services

WTP increases with household income in line with our expectations and with the results reported by Beltran and Rojas (1996), Santagata and Signorello (2000), and Mazzanti (2001). However, some authors (Smith et al., 1983; Ashworth and Johnson, 1996) also mention the possibility of a negative correlation with income when considering leisure activities, such the visit to a museum, because those who have higher labour income also face higher opportunity costs to visit the site.

Table 12.4 WTP depending on household income (552 observations)

Variables	Coeff.	Std.Dev.	t-stat	*	**
HOURS	0.6614	0.0748	8.842	4.17	6.93
AUDIO	0.4607	0.0635	7.253	2.91	5.14
TOURS	0.8181	0.0678	2.070	5.16	8.57
BAR	-0.2118	0.0737	2.875	1.34	-2.22
EXHIBIT	0.2835	0.0734	3.860	1.79	2.97
EVENT	0.4700	0.0767	6.128	2.97	4.92
LAB	0.7181	0.0809	8.879	4.53	7.52
AUDIOV	0.4906	0.0740	6.631	3.01	5.14
DOCUM	0.4082	0.0623	6.549	2.58	4.29
Marginal utility of income				0.158	0.095

* marginal WTP in euros per each service when income = €20,658
** marginal WTP in euro per each service when income = €46,481

Table 12.5 shows the estimates obtained by dividing the whole sample into subsets according to the different levels of education. It is interesting to note that respondents with a level of education inferior to college degree do not feel the presence of a café in the archaeological area as a negative feature, as shown by a coefficient which is no longer significant.

Table 12.5 WTP depending on education

Variables	Subsample with level of education >= College degree (282 observations)				Subsample with level of education < College degree (270 observations)			
	Coeff.	Std.Dev.	t-stat	*	Coeff.	Std.Dev.	t-stat	*
HOURS	0.8229	0.1075	7.654	5.00	0.5668	0.1056	5.368	3.97
AUDIO	0.5878	0.0882	6.667	3.57	0.3687	0.1075	3.429	2.58
TOURS	0.9299	0.0947	9.817	5.65	0.7198	0.0968	7.438	5.04
BAR	-0.3883	0.1023	-3.794	-2.36	-0.0330	0.1010	-0.327	-0.23
EXHIBIT	0.2445	0.1133	2.158	1.48	0.3404	0.1187	2.867	2.38
EVENT	0.5601	0.1093	5.126	3.40	0.3943	0.1083	3.640	2.76
LAB	1.0180	0.1293	7.870	6.18	0.4520	0.1105	4.092	3.17
AUDIOV	0.5460	0.1006	5.428	3.32	0.3867	0.1246	3.103	2.71
DOCUM	0.4974	0.0971	5.089	3.02	0.2994	0.0926	3.232	2.09
Marginal utility of income			0.164					0.143

* marginal WTP in euros per each service when income = €20,658

At the same time, the order of preference for the different services changes, since for people holding a degree the lab is the most preferred service, while the respondents with no college degree prefer longer opening hours and guided tours. In general, in our sample the WTP increases with the level of education, as is also reported in other studies, e.g. Beltran and Rojas (1996) and Mazzanti (2001).

Table 12.6 shows the estimates obtained by dividing the sample into two subsets corresponding to two age levels: less or more than 33 years. In both cases, guided tours represent the most preferred service. However, we see that people who fall in the older group are more willing to pay for an increase in opening hours, whilst the younger ones prefer a lab and audiovisuals.

The latter group also shows a coefficient for the variable BAR that is no longer significant, while the WTP is higher for the older group, probably because older people are more likely to be earners. Our estimates confirm the results reported by Mazzanti (2001) and Morey and Rossman (2002), while an opposite result can be found in Santagata and Signorello (2000).

Table 12.6 WTP depending on age

Variables	Subsample with age<= 33 years (126 observations)				Subsample with age >33 years (426 observations)			
	Coeff.	Std.Dev.	t-stat	*	Coeff.	Std.Dev.	t-stat	*
HOURS	0.1792	0.2221	0.807	0.93	0.7633	0.0802	9.514	5.36
AUDIO	0.4344	0.1491	2.914	2.26	0.4687	0.0754	6.219	3.29
TOURS	1.0598	0.1510	7.018	5.50	0.7764	0.0741	10.479	4.45
BAR	-0.2436	0.1561	-1.560	-1.26	-0.1940	0.0802	-2.419	-1.36
EXHIBIT	0.2827	0.1620	1.745	1.47	0.3006	0.0916	3.283	2.11
EVENT	0.4313	0.1508	2.860	2.26	0.4861	0.0857	5.669	3.41
LAB	0.7446	0.1801	3.886	3.84	0.7198	0.0938	7.673	5.05
AUDIOV	0.6873	0.1740	3.713	3.55	0.4600	0.0874	5.266	3.23
DOCUM.	0.3314	0.1564	2.118	1.82	0.4339	0.0737	5.887	3.04
Marginal utility of income			0.192					0.143

* marginal WTP in euros per each service when income = €20,658

Table 12.7 shows that no significant differences can be found between the preferences expressed by residents in the Campania Region and residents elsewhere, except for those concerning the café within the archaeological area, which is perceived negatively only by residents. A stronger preference for concerts and performances is found among residents, probably because of their ease of accessing the site throughout the year.

Finally, Table 12.8 shows estimates for two groups of male and female respondents, reported on the grounds that they may have different preferences. The results show that women have a higher WTP for performances and other events than men, while men have a higher WTP for guided tours than women.

12.5 Concluding remarks

One of the first issues to be resolved in order to find optimal policies for the management of museums, and of cultural goods in general, is the definition of the main and most desirable output, the final goal of the policy, whether it is conservation, education or something else. Different 'stakeholders' would probably have different perceptions of what the most desirable output is. An 'intellectual' might perceive art as belonging to an elite whose principal objective and purpose is to preserve the work of art for future generations, or even for its own sake. Someone with a more social vision of art and cultural heritage might be more interested in promoting the knowledge of this archaeological site, maybe envisaging free access. A local administrator might prefer a policy aimed at attracting more tourism, hence encouraging all the services that may complement tourism, such as the more leisure-oriented ones. An optimal policy should account for all the different positions, including that of the general public.

Table 12.7 WTP depending on residence

Variables	Subsample of residents in Campania (190 observations)				Subsample of non residents (362 observations)			
	Coeff.	Std.Dev.	t-stat	*	Coeff.	Std.Dey.	t-stat	*
HOURS	0.7241	0.1658	4.368	3.98	0.6481	0.0843	7.688	4.29
AUDIO	0.5040	0.1082	4.658	2.77	0.4396	0.0861	5.107	2.91
TOURS	0.9313	0.1259	7.398	5.12	0.7887	0.0799	9.872	5.22
BAR	-0.3453	0.1203	-2.871	-1.90	-0.1265	0.0879	-1.439	0.84
EXHIBIT	0.3640	0.1343	2.710	2.00	0.2817	0.1035	2.721	1.86
EVENT	0.6623	0.1274	5.200	3.64	0.3791	0.0966	3.925	2.56
LAB	0.7188	0.1575	4.564	3.95	0.6701	0.0983	6.815	4.44
AUDIOV	0.6940	0.1469	4.723	3.81	0.3902	0.0931	4.190	2.58
DOCUM	0.5664	0.1170	4.480	3.11	0.3230	0.0812	3.978	2.14
Marginal utility of income			0.182					0.151

* marginal WTP in euros per each service when income = €20,658

Table 12.8 WTP depending on gender

Variables	Subsample of women (190 observations)				Subsample of men (362 observations)			
	Coeff.	Std.Dev.	t-stat	*	Coeff.	Std.Dev.	t-stat	*
HOURS	0.6575	0.1090	6.034	4.26	0.6471	0.1025	6.310	3.71
AUDIO	0.4334	0.1020	4.247	2.81	0.4928	0.0915	5.385	2.83
TOURS	0.9333	0.1031	9.050	6.05	0.7843	0.0893	8.781	4.50
BAR	-0.2383	0.1003	-2.376	-1.54	-0.2039	0.0995	-2.049	1.17
EXHIBIT	0.2092	0.1171	1.787	1.36	0.3808	0.1138	3.345	2.19
EVENT	0.8147	0.1119	7.283	5.28	0.2206	0.1091	2.023	2.47
LAB	0.6066	0.1301	4.661	3.93	0.7782	0.1147	6.784	4.47
AUDIOV	0.7180	0.1341	5.356	4.65	0.3862	0.0983	3.928	2.22
DOCUM	0.4272	0.0910	4.696	2.77	0.3815	0.0975	3.913	2.19
Marginal utility of income			0.154					0.174

* marginal WTP in euros per each service when income = €20,658

This study analyzes visitors' preferences for alternative museum services. The results seem interesting for the development of new management policies for the Temples of Paestum, and appear to confirm the potential that stated preference valuation techniques, such as the conjoint analysis approach used in our study, have for these purposes. In particular, we find that the most preferred services are those which improve the accessibility and the understanding of the site, including its museum (longer opening hours, guided tours), followed by educational services such as a children's lab. Our results confirm that the main reason motivating people to visit cultural sites is the desire to 'learn something', as is also argued by Verbeke and van Rekom (1996).

The WTP to gain access to the site increases with age, education, and income, confirming previous results. The majority of respondents show no interest towards the transformation of this cultural site into a sort of entertainment centre, with the organization of performances or special events, and the creation of a café within the archaeological remains is perceived negatively. However, people with a lower level of education, the youngest, and tourists non-resident in the region do seem interested in the creation of the café.

In sum, we can say that our results show a preference for a management policy oriented towards improvement in accessibility, linked with educational and pedagogical purposes. This confirms a trend shown in many European museums, where the principal focus is on the exhibited good, more than on the other services that are considered ancillary, and sometimes separated from the museum.

From our results, we feel able to support our initial statement that conjoint analysis methodology appears particularly suitable to assess management options. Nonetheless, the potential and limits of this technique need to be fully explored. Further research is needed to test how stated preference techniques might be used for management purposes of cultural sites and their services. Since cultural tourism, with its positive and negative impacts on cities' economies, is becoming a relevant and debated issue, more studies are needed in this area, aiming to assess the non-market externalities brought by tourists' presences in historic sites, as well as their level of satisfaction. How tourists value the attributes that make up their visit experience and how residents value the impacts that tourism oriented management strategies bring upon their quality of life, are among the issues that future valuation studies and applications of conjoint analysis might need to focus on.

Acknowledgements

The authors wish to thank Giuliana Tocco, Superintendent for the Temples of Paestum, for her cooperation and insights during the scenario development's phase. Acknowledgments are also due to Ken Willis and Anna Alberini for their statistical advice on choice attributes' combination. Final thanks are due to Adalgiso Amendola, Dipartimento Scienze Economiche Universita' degli Studi di Salerno, coordinator of the project, funded by Regione Campania (L. 41/1994), where this study originated from. All errors remain solely the authors' responsibility.

References

Alberini, A., P. Riganti and A. Longo (2003) 'Can People Value the Aesthetic and Use Services of Urban Sites? Evidence from a Survey of Belfast Residents', *Journal of Cultural Economics*, 27(3–4): 193–213.

Ashworth, J. and J. Johnson (1996), 'Sources of "Value for Money" for Museum Visitors: Some Survey Evidence', *Journal of Cultural Economics*, 20: 67–83.

Beltran, E. and M. Rojas (1996), 'Diversified Funding Methods in Mexican Archeology', *Annals of Tourism Research*, 23(2): 463–78.

Feldstein, M. (1991), *The Economics of Art Museums*, University of Chicago Press, Chicago.

Frey, B.S. (1994), 'Cultural Economics and Museum Behaviour', *Scottish Journal of Political Economy*, 41: 325–35.

Frey, B.S. and W.W. Pommerehne (1989), *Museums and Markets: Explorations in the Economics of Arts*, Blackwell, Oxford.

Hanley, N., R.E. Wright and W. Adamowicz (1998), 'Using Choice Experiments to Value the Environment: Design Issues, Current Experience and Future Prospects', *Environmental and Resource Economics*, 11: 413–28.

Harrison, J. (1997), 'Museums and Touristic Expectations', *Annals of Tourism Research*, 24(1): 23–40.

Healey, P. (1998), 'Collaborative Planning in a Stakeholder Society', *Town Planning Review*, 69(1): 1–20.

Jackson, R. (1988), 'A Museum Cost Function', *Journal of Cultural Economics*, 12: 41–50.

Johnson, P.S. and R.B. Thomas (1998), 'The Economics of Museums: A Research Perspective', *Journal of Cultural Economics*, 22: 75–85.

Luxen, J.L. (2000), Introductory statement at the conference on 'Cultural Heritage Management and Urban Development: Challenge and Opportunity', *Conference Proceedings*, 5-7 July, Beijing.

Mazzanti, M. (2001), 'Discrete Choice Models and Valuation Experiments: An Application to Cultural Heritage', *Working Paper*, SIEP 75.

Mitchell, R. and R. Carson (1989), *Using Surveys to Value Public Goods: The Contingent Valuation Method. Resources for the Future*, Washington DC.

McFadden, D. (1973), 'Conditional Logit Analysis of Qualitative Choice Behavior', in P. Zarembka (ed.), *Frontiers of Econometrics*, Academic Press, New York, pp. 105–142.

Morey, E. and K.G. Rossman (2002), 'Using Stated-Preference Questions to Investigate Variation in Willingness to Pay for Preserving Marble Monuments: Classic Heterogeneity and Random Parameters', *Working Paper*, Economics Department, University of Colorado at Boulder, Colorado, 10 January 2002.

Navrud, S. and R. Ready (2002), *Valuing Cultural Heritage*, Edward Elgar Publishing, Cheltenham.

Noonan, D. (2002), 'Contingent Valuation Studies in the Arts and Culture: An Annotated Bibliography', *Working Paper*, The Cultural Policy Center, University of Chicago, Chicago.

Santagata, W. and G. Signorello (2000), 'Contingent Valuation of a Cultural Public Good and Policy Design: The Case of "Napoli Musei aperti"', *Journal of Cultural Economics*, 24(3): 181–204.

Scarpa, R., G. Sirchia and M. Bravi (1998), 'Kernel vs. Logit Modeling of Single Bounded CV Responses: Valuing Access to Architectural and Visual Arts Heritage in Italy', in R. Bishop and D. Romano (eds), *Environmental Resource Valuation: Applications of Contingent Valuation Method in Italy*, Kluwer Publishers, Boston.

Silbeberg, T. (1995), 'Cultural Tourism and Business Opportunities for Museums and Heritage Sites', *Tourism Management*, 16(5): 361–5.

Smith, V.K., W. Desvouges and M.P. McGivney (1983), 'The Opportunity Cost of Travel Time in Recreation Demand Models', *Land Economics*, 59.

Thompson, E., M. Berger, G. Blomquist and S. Allen (2003), 'Valuing the Arts: A Contingent Valuation Approach', *Conference on Contingent Valuation of Culture*, 1-2 February, Chicago.

Verbeke, J. and M. van Rekom (1996), 'Scanning Museum Visitors', *Annals of Tourism Research*, 23(2): 364–75.

PART III
Policy Strategies on Tourism

Chapter 13

The Importance of Friends and Relations in Tourist Behaviour: A Case Study on Heterogeneity in Surinam

Pauline Poel, Enno Masurel and Peter Nijkamp

13.1 Scope and aim

The outreach of tourism as a modern economic sector is on a rising edge. Since the 1960s, tourism has expanded into all corners of the earth (Theobald, 2004; Theuns, 2002). Because of the growing importance of tourism and its potential economic value for a country or region, tourism has become a popular object of study. There is, however, one category of tourists who have as their main purpose 'visiting friends and relatives' (VFR) which has been neglected, most likely because they are assumed to have a secondary status when measured in economic terms. Yet more and more researchers are questioning whether the economic contribution of VFR tourists is really insignificant. After a long period of overlooking the VFR market in most international tourism studies, the VFR market is increasingly becoming a subject of research. For example, Jackson (1990) demonstrated that the extent of VFR tourism to Australia is underestimated: many visitors classified as holidaymakers actually spent much of their time with friends and relatives. If Jackson is right, the VFR market would be much bigger than has previously been thought.

Another question is whether the VFR market has many common features. Can we speak of one homogeneous VFR market? Moscardo et al. (2000) were some of the first to question this homogeneity. They demonstrated the heterogeneity within the VFR market and created a typology of VFR tourists. A heterogeneity within the VFR market could have consequences for the assessment of the economic value of VFR tourism. The principal objective of our study is to analyze the heterogeneity within the VFR tourism market in economic terms. The main question to be answered in this study is therefore: Is a typology of VFR tourists economically relevant? In other words, can we distinguish different types of VFR tourists who differ from each other in terms of their economic impact?

The first part of the chapter (Sections 13.2 to 13.5) provides an overview of the literature and the outcomes of earlier research on the heterogeneity within the VFR tourism market. The second part of the chapter (Sections 13.6 and 13.7) then goes on to test empirically the heterogeneity within the VFR market in Surinam. The distinction of different types of VFR tourists is based on two typology factors proposed earlier by Moscardo et al. (2000). We will also address two new factors, viz. differences in the expenditures of non-ethnic and ethnic VFR tourists, as well as differences between first- and successive-generation VFR tourists. Consequently, this study provides useful and new knowledge on VFR tourism, while it is also one of the first VFR studies conducted in a developing country. The chapter closes (in Section 13.8) with a number of conclusions and recommendations for further research.

13.2 A classification of tourism

There are many forms of tourism, ranging from short city trips to world tours, from vacations in 'all-inclusive resorts' in popular places like Turkey to expeditions to almost unknown parts of the world. The various types of tourists do differ in their behaviour. These behavioural differences may result in differences in spending patterns, which influence the economic impact. Therefore, it is useful to make a classification of tourism.

A first distinction is made by the United Nations Conference on International Travel and Tourism at Rome in 1963 between an *overnight tourist* and a *same-day visitor* (UN, 1990), as defined below:

- Overnight tourist: temporary visitor staying at least 24 hours in the country visited, whose journey purpose can be classified under one of the following headings:
  - o Leisure (recreation, holiday, health, study, religion and sport);
  - o Business, family, mission, meeting.
- Same-day visitor: temporary visitor staying less than 24 hours in the country visited (including tourists on cruise ships). Also known as 'excursionists'.

A second distinction can be made between domestic tourists and international tourists. Another frequently encountered classification is by purpose of visit. The United Nations and WTO (2000) proposed the following classification:

- leisure, recreation and holidays;
- visiting friends and relatives (VFR);
- business and professional;
- health treatment;
- religion/pilgrimages;
- other.

This classification is used in many tourism impact studies. Most times, the focus of these studies is on leisure tourists because, according to this classification, they are the only ones who undertake strictly tourist activities.

However, it may be possible that tourists combine a number of different purposes in the course of their visit. The description of the activities of the different types of tourists by McIntosh et al. (1995) is therefore more realistic. They took into account the possibility of multiple-purpose visits, and made a distinction between primary and secondary activities. In Table 13.1, the primary and secondary activities of the different types of tourists are presented.

Table 13.1 The primary and secondary activities of the different types of tourists

Tourism markets	Primary activities	Secondary activities
Business	Consultations; Conventions; Inspections	Dining out; Recreation; Shopping; Sightseeing; VFR
VFR	Socializing, Dining at home; Entertainment	Dining out; Recreation; Shopping; Sightseeing; Urban Entertainment
Other personal business	Shopping; Visiting lawyer; Medical appointment	Dining out; VFR
Pleasure	Recreation; Sightseeing; Dining out	VFR; Convention; Business; Shopping

Source: McIntosh et al. (1995).

This table makes it clear that the different tourism markets, such as business, VFR, personal business and pleasure are not homogeneous markets because of the number of different purposes combined in one visit. Moreover, for the same reason it is difficult to distinguish the different types of tourists from each other. In many instances, the categorization was therefore not made by the researchers but by the tourists themselves, who had to fill in the primary purpose of their visit on their arrival card.

Before discussing the heterogeneity within the VFR market, it is first necessary to describe the characteristics of the VFR market in order to come to a better understanding of this market.

13.3 VFR tourism

In 1995 a special issue of the *Journal of Tourism Studies* was devoted to the VFR market, with the aim of stimulating tourism-researchers to take into account VFR tourism in their analyses (Morrison and O'Leary, 1995). Till then, the VFR market was largely ignored. According to Jackson (1990), this lack of interest was because VFR tourism was assumed to constitute only a small percentage of total overseas visits, and, because VFR tourists do not utilize tourist facilities, the formally constituted tourism industry has little interest in them. Maybe the most important reason for this lack of interest is the assumed low level of spending by VFR tourists (Seaton and Palmer, 1997; Lehto et al., 2001), which make them not very interesting from an economic perspective.

However, more and more people are questioning the validity of these reasons and now include VFR tourism in their analyses. One reason for the growing interest in VFR tourism is the size of the market; in some destinations VFR tourism is even the principal source of tourists (Seaton and Palmer, 1997). VFR is an important segment of leisure travel not only in industrialized nations, but also in some developing countries (Müri and Sägesser, 2003). Jackson (1990) studied VFR tourism in Australia and, after studying only the visitor numbers, he was able to conclude that the significance of VFR tourism is greatly underestimated. The average length of stay of a VFR tourist is twice as long as the average length of stay of a leisure tourist, and nearly three times as long as the average length of stay of a business tourist. Furthermore, the significance of the VFR market is greatly underestimated because of the limited definition of a VFR tourist. Also in this case, someone is classified as a VFR tourist when he/she gave 'visiting friends and relatives' as the main purpose of his/her visit. So the utilized classification of types of tourist depends largely upon self-assessment by the tourist when filling in arrival cards for immigration purposes. Whilst straightforward holidaymaking tourists are very unlikely to classify themselves as VFR tourists, persons who are VFR tourists could state that they were holidaymakers.

There is not much data available considering the size of the VFR market in terms of number of arrivals. However, we found a table in an OECD study that presents the number of tourists who arrived at their holiday destination as 'arrival by purpose of visit'. This table shows that VFR tourism, measured in arrivals per year, is quite big, varying from almost 19 per cent to almost 30 per cent of total arrivals (OECD, 1997). The share of the World Tourism Organization (WTO) rest-category (VFR tourism, pilgrimages, health treatment, and other) has been rising, especially in the last ten years (WTO 2001). VFR tourism is not only a well-known phenomenon in former colonized countries. VFR tourism has also grown alongside the development of international migration of labour in more recent years. An example of a country where VFR tourism has grown as a consequence of the migration of residents to Western Europe is Morocco. Migration generates these VFR tourism flows either because migrants may become poles of tourist flows, in the sense that friends and relatives come to visit them, or because migrants themselves become tourists when returning to visit friends and relatives in their areas of origin (Williams and Hall, 2000). Jackson (1990) confirms this positive relation between migration and VFR tourism. He concluded for Australia that the volume of total VFR tourism, both inward and outward, is reasonably closely and significantly associated with the size of different migrant groups in Australia and their period of residence in the country (Jackson, 1990). This positive relation implies that, when migration flows keep on growing, VFR tourism flows will be growing as well.

When compared with other pleasure tourists, the VFR tourist has some different characteristics in choices of travel times and destinations, travel information search and trip planning behaviour, accommodation use, spending patterns and trip activities (Hu and Morrison, 2002). We will provide a few examples below.

Long haul international VFR tourists stay longer at their destination (Seaton and Tagg, 1995; Yuan et al., 1995). However, Seaton et al. (1997) and Müri and Sägesser (2003) found contrary results for domestic VFR tourism. These VFR trips were shorter in comparison with all the other trips. The difference between domestic and international VFR patterns may account for this discrepancy (Seaton and Palmer, 1997). However, Hu and Morrison (2002) found that domestic VFR tourists in the US and Canada tend to stay longer at their destination. The same result is found by Jackson (1990) for the Australian domestic VFR tourists and by Morrison et al. (1995) for the American domestic VFR tourists. This discrepancy could be caused by the longer travel distances in larger countries like the US, Canada and Australia in comparison with the smaller countries like the UK and Switzerland.

Furthermore, both Hu and Morrison (2002) and Noordewier (2001) found that VFR tourists were more likely to make trips in the off-season than other types of tourists. Because of their relationship with friends and relatives VFR tourists are more likely to be repeat visitors to many destinations (Noordewier, 2001; Meis et al., 1995).

Although it is true that most VFR tourists stay with their friends and relatives, many studies have pointed out that a considerable part of the VFR tourists do use commercial accommodation (Noordewier, 2001; Lehto et al., 2000; Braunlich and Nadkarni, 1995; Morrison et al., 1995). Besides the use of commercial accommodation, other tourism-related facilities were appreciated and used by Dutch VFR tourists to Canada (Yuan et al., 1995). It appeared that they were more likely to enjoy those facilities closer to those urban areas, and city activities that are accessible to the guest and the host group. This is confirmed by the study of Seaton et al. (1997), which found that VFR tourists tend to stay more in urban regions than in rural and seaside destinations.

And finally, international VFR tourists appear to have significant expenditures on food and beverages and entertainment (Lehto et al., 2001; Morrison et al, 1995).

13.4 Heterogeneity within the VFR market

A common approach in studying the VFR market has been to regard it as one homogeneous market without significantly different component market segments (Morrison et al., 1995). However, more and more people are questioning this approach.

Moscardo et al. (2000) questioned the homogeneity of the VFR market and proposed a typology of VFR travel, which is based on earlier research. This initial typology, slightly modified, is represented in Table 13.2.

Table 13.2 A typology of VFR tourists

Sector	Scope	Effort	Accommodation used	Focus of visit
Visiting friends and relatives as	Domestic	Short haul	Non-comm. VFR (staying only with friends and relatives)	VF (visiting friends), VR (visiting relatives), VFVR (visiting friends and relatives)
1. main purpose or 2. as an activity			Comm. VFR (accommodated at least one night in comm. accom.)	VF, VR, VFVR
		Long Haul	Non-comm. VFR	VF, VR, VFVR
			Comm. VFR	VF, VR, VFVR
	International	Short haul	Non-comm. VFR	VF, VR, VFVR
			Comm. VFR	VF, VR, VFVR
		Long haul	Non-comm. VFR	VF, VR, VFVR
			Comm. VFR	VF, VR, VFVR

Source: Moscardo et al. (2000).

As already said in Section 13.1, two factors of the typology of Moscardo et al. (2000) and two new factors will be used to test empirically the heterogeneity within the VFR tourism market. This section provides an overview of the outcomes of earlier research on the heterogeneity within the VFR tourism market using these four typology factors.

13.4.1 The factor sector, VFR as main purpose versus VFR as an activity

When VFR is the sole purpose of visit, the whole travel experience might be focussed on social obligations. On the other hand, when VFR is just an activity for a tourist he/she might participate in a range of tourist activities (Moscardo et al.,

2000). VFR tourists with VFR just as an activity are more likely to stay in commercial accommodation, more likely to participate in tourist activities, and spend more money than VFR tourists with VFR as their main purpose (Moscardo et al., 2000; Lehto et al., 2001).

13.4.2 The factor accommodation used, commercial versus non-commercial VFR

VFR tourists who stay at least one night in commercial accommodation tend to spend more on food and beverages, transportation, gifts and souvenir shopping, and entertainment than VFR tourists who stay only at the houses of friends and relatives (Lehto et al., 2001).

13.4.3 The factor migration, ethnic versus non-ethnic VFR

A factor that is not included in the typology but that could have consequences for the heterogeneity within the VFR market is migration. Some authors see migrants who return to visit friends and relatives in their areas of origin as a separate form of tourism: namely, ethnic tourism. Ostrowski (1991) defined ethnic tourism as follows: 'travel to an ancestral home without the intention of permanent settlement, emigration or re-emigration, or undertaking temporary paid work.' This travel could be motivated by a desire to delve into family history through travel to the relevant country. It might, or might not, involve actual staying with the family (King, 1994). Furthermore, ethnic tourism is particularly important in the Third World (Wood, 1998). Earlier studies have demonstrated the importance of remittances for developing countries with high emigration rates (Adams and Page, 2003; Osaki, 2003). When visiting their country of origin, it is likely that ethnic VFR tourists bring gifts with them and contribute to the economy of this developing country. It is therefore interesting to study whether ethnic tourists are more likely to bring gifts for their friends and relatives than non-ethnic VFR tourists.

13.4.4 The factor generation, first-generation versus successive-generation migrants

According to the above-mentioned definition, ethnic tourists are both the first migrants and the successive generations of migrants. While the main purpose of visit for the first generation of migrants is probably VFR, for successive generations the search for 'their roots' can be the main purpose of their visit (King, 1994). As a consequence, successive generations could be more interested in exploring the country of origin than in just visiting friends and relatives. This would imply that this group is more likely to undertake excursions. Moreover, it is plausible that successive generations have less contact with the relatives in their country of origin in comparison with the first migrants. A possible consequence could be that successive generations are less likely to stay with their friends and relatives and make more use of commercial accommodation. Unfortunately, no earlier research was found which studied the differences in behaviour between first-generation migrants and second-generation migrants.

13.5 The economic contribution of VFR tourism

Many economic impact analyses are carried out for different purposes. When measuring the economic contribution, it is necessary to trace the flows of spending associated with tourism activity in a region in order to identify changes in sales, tax revenues, income, and jobs due to tourism activity (Frechtling, 1994). In general, there are three types of effects of tourism (Liu et al., 1984):

1. Direct effects: these effects account for the income generated as tourists make purchases from the tourist-related businesses.
2. Indirect effects: these effects occur as the tourist-related businesses where tourists make local purchases from all other enterprises in the studied region.
3. Induced effects: the additional earnings for the people employed at the tourist-related businesses or their supplying businesses give rise to the household incomes and the level of spending of these households.

When measuring the economic impact from a demand-side perspective, visitor expenditure is the basic complement. The visitor expenditure can be divided into three components (UN and WTO, 2000):

1. *all* consumption expenditure made *during* the trip by a visitor;
2. consumption expenditure made *before* the trip by a visitor in goods and services *necessary* for the preparation and undertaking of the trip;
3. consumption expenditure made *after* the trip by a visitor on those goods and services whose use is clearly *related* to the trip.

In order to know more about the visitor and his/her economic impact, expenditure research is not limited to the question of how much visitors have spent in total but also covers how their expenditures are distributed among different categories. Another reason for the importance of dividing the expenses into different categories is the difference in leakage rates among the different sectors within the tourism industry.

Because different types of tourist may differ in terms of their spending patterns, it is likely that the size of the leakages differs between the different types of tourists. Hotels, for example, need more capital investment than smaller tourism establishments, and are therefore more likely to attract foreign capital and, as a consequence, have a greater leakage. It is likely that the leakages are higher for tourists whose expenditure is largely directed to commercial accommodation than for tourists whose expenditure is mainly directed to smaller tourism establishments or to non-tourist establishments (Jackson, 1990).

A well-known classification of visitor expenditures is that of the United Nations and the WTO (2000). They distinguish the following categories: package travel; accommodation; food and drinks; transport; recreation, culture and sporting activities; shopping and other activities. According to their recommendations for tourism statistics, cash given to relatives or friends and donations to institutions should be left out of the analysis. However, the exclusion of cash given to relatives or friends during a holiday trip could have some serious consequences for the level of expenditures of VFR tourists in general, and of VFR tourists with an ethnic connection in particular. This is demonstrated by Asiedu (2003, p. 12) who did a study of tourism in Ghana. It appeared that tourists spent 20 per cent of their total expenses on 'other incidentals, i.e. contributions to development funds, funeral expenses'.

Jackson (1990) studied the VFR market in Australia and concluded that, although the VFR tourist spent less than other types of tourists, he/she spent more in smaller tourism establishments or in non-tourism establishments. This finding is confirmed by the studies of Stynes (2001) and Seaton and Palmer (1997). The leakages are lower for a VFR tourist whose expenditure is more likely to be directed to smaller tourism establishments or to non-tourism establishments than they are for VFR tourists whose expenditure is largely directed to commercial accommodation (Jackson, 1990). Liu et al. (1984) determined the tourist-income multipliers for various groups of tourists to Turkey, and reached the same conclusion as Jackson about the leakages.

The assumption that VFR tourists do not use commercial accommodation is not true for all VFR tourists. The research of Braunlich and Nadkarni (1995) demonstrated that the VFR market is of considerable importance to the hotel industry in the East North Central census region of the USA. The VFR tourists had a significantly lower expenditure per hotel room night than pleasure and business tourists. However, this lower daily expenditure level is partly compensated by the longer hotel stays of VFR tourists in comparison with pleasure and business tourists. Furthermore, there are VFR tourists who do use commercial accommodation (Braunlich and Nadkarni, 1995), and VFR tourists are often repeat visitors and therefore spend more during their travel lifecycles (Meis et al., 1995).

Only a few studies have been carried out where different types of VFR tourists were compared with each other in terms of their expenditures. The most important findings were:

- VFR tourists who reported VFR as their main travel purpose spent less on accommodation than tourists for whom VFR was a secondary purpose (Lehto et al., 2001).
- Tourists who stayed at least one night in commercial accommodation spent significantly more in total, as well as for transportation, food and beverages,

lodging, and entertainment, than tourists who stayed in non-commercial accommodation (Lehto et al., 2001).

- Visiting Friends (VF) tourists spent more on entertainment and drinks than VFR tourists. VF tourists were: slightly less likely to spend money on transport than Visiting Relatives (VR) tourists; much less likely to buy souvenirs or presents than VR tourists; and much less likely to spend money on shopping than both VR tourists and VFR tourists (Seaton and Tagg, 1995).

- Unfortunately, no research material is found where ethnic VFR tourists are compared with non-ethnic VFR tourists in terms of their spending patterns. One of the few economic impact analyses in which ethnic tourists are distinguished as a separate category are the studies of Liu et al. (1984) and Gamage and King (1999). They both found that ethnic tourists spend more on retail goods and less on hotels and restaurants. The lower spending on hotels by the ethnic tourists is not surprising. Because they have ethnic relations, it is very plausible that they are staying with their relatives instead of in commercial accommodation. However, the lower spending on restaurants seems to contradict the results of general VFR tourism research that found that VFR tourists had significant expenditures in the catering industry. The difference between ethnic VFR and non-ethnic VFR might be the reason for this contradiction.

We will now use Surinam as our empirical test case.

13.6 Surinam in a nutshell

Surinam is the smallest independent country on the South American continent. Surinam borders French Guiana in the west, (British) Guyana in the east, Brazil in the south and the Atlantic Ocean in the north. Surinam has an area of 163,270sq km, which is almost 4 times the area of the Netherlands. The main part of the country is covered with tropical rainforest with a great diversity of flora and fauna.

Surinam has a relatively small population of around 435,449 (July 2003, estimate). Most of the people (around 75 per cent) live in Paramaribo. Almost nobody lives in the interior, which is quite uninhabitable because of the dense forest. Despite the high birthrate (18.87 births/1,000 population), the size of the population has hardly changed. This is primarily caused by the high emigration rate (8.81 migrants/1,000 population). One of the consequences of this high emigration rate is the 'brain drain'. It is especially highly-educated people who leave the country.

The Surinamese population is a mixture of various cultures. The following ethnic groups can be distinguished: in the population 37 per cent are Hindustani (their ancestors emigrated from northern India in the latter part of the nineteenth century), 31 per cent Creole (mixed white and black), 15 per cent Javanese, 10 per cent Maroons (their African ancestors were brought to the country in the seventeenth and eighteenth centuries as slaves and escaped to the interior), 2 per cent Amerindians, 2 per cent Chinese, 1 per cent white, and 2 per cent other ethnic groups. Because of the population mix, many religions are represented in the Surinamese culture: Hindu 27.4 per cent, Muslim 19.6 per cent, Roman Catholic 22.8 per cent, Protestant 25.2 per cent (predominantly Moravian), and indigenous beliefs 5 per cent. This diversity makes the country very interesting from the tourism perspective but it also makes the country very complex because the diversity in cultures has great influence on the social and political system, which has consequences for the economy (http://www.cia.gov/cia/publications/factbook/geos/ns.html).

13.7 Case study on Surinam

The purpose of the study is to determine whether the typology factors, 'sector', 'accommodation used', 'migration' and 'generation' influence the spending patterns of VFR tourists. As it appears that different types of VFR tourists each have a different spending pattern, we can say that a typology of VFR tourists is useful for economic purposes. The study focusses on the VFR tourists resident in the Netherlands, travelling by air and departing from the JAP airport of Surinam.

13.7.1 Methodology

In the period December 2003–February 2004, a survey was conducted at the JAP airport.[1] All departing visitors (not resident in Surinam) were asked to fill in a questionnaire. In total, 926 people were approached, that is about 2.3 per cent[2] of the annual VFR tourists in Surinam. 795 respondents cooperated with the survey and filled in the questionnaire (2.0 per cent of the annual VFR tourists in Surinam).

Because the study puts the emphasis on the spending pattern of the VFR tourists, most questions are related to the various expenditures of the tourists. Respondents had to distribute their total expenditures across 18 categories. In contradiction with the recommendations of the WTO and UN, gifts to friends and relatives are included (see Section 13.5). Unfortunately, not all the questionnaires could be used for analysis. Some respondents fell outside the target group because they were not Dutch or were tourists in transit. After exclusion of the questionnaires filled in by the latter groups, 745 questionnaires were left. Of those 555 respondents filled in the expenditure-related questions, which were selected for the analysis.[3]

---

[1]   The international airport of Surinam, also known as Zanderij.

[2]   The size of the VFR tourism market was estimated as 39,439 travellers per year (926 / 39,439) * 100 per cent equals 2.3 per cent (based on the figures of the CTO and STS, 2001).

[3]   The socio-demographics and trip-related characteristics of the 745 and the 555 questionnaires were compared with each other and no significant differences were found.

In most studies, a VFR tourist is defined as someone who filled in 'visiting friends and relatives' as the main purpose of his/her visit. In this study, a broader definition will be used. A VFR tourist is defined as someone:

- who filled in VFR as his/her main purpose of visit; and/or
- who stayed at least one night at the house of his relatives and friends.

When this definition is used, 480 respondents can be considered as VFR tourists. Someone who filled in another main purpose of visit than VFR, but stayed at least one night at the house of his/her friends and/or relatives is considered as a VFR tourist with VFR as an activity.

Using the four typology factors, the respondents were clustered into 8 different categories (see Table 13.3).

Table 13.3   The eight different categories of VFR tourists (in brackets the number of respondents)

VFR as main purpose (365)	VFR as activity (115)
Commercial VFR (VFR tourists who stay at least one night in commercial accommodation) (39)	Non-commercial VFR (VFR tourists who stay at the houses of friends or relatives or in other non-commercial accommodation) (440)
Ethnic VFR (first- and successive-generation VFR) (414)	Non-ethnic VFR (VFR tourists who are not born in Surinam, neither are their (grand)parents) (66)
First generation VFR (VFR tourists who are born in Surinam and migrated to the Netherlands) (341)	Successive generation VFR (VFR tourists who have at least one (grand)parent born in Surinam) (73)

In this study, three types of analysis are conducted. First, the characteristics of the VFR tourists in Surinam will be compared with the characteristics of the VFR tourists found in the literature, in order to determine whether VFR tourism in Surinam has the same characteristics as VFR tourism described in the literature. Second, statistical comparisons will be made using chi-square tests for categorical variables and t-tests for continuous variables, in order to verify whether statistical differences exist between the different types of VFR tourists in terms of socio-demographic and trip-related characteristics. Third, statistical comparisons will be made using t-tests to verify whether statistical differences exist between the different types of VFR tourists in terms of their expenditures. Differences are

---

Therefore, it can be assumed that the selection had no consequences for the outcomes of this study.

considered as significant when the significance levels are lower than 0.05, and they are considered as indicative when the significance levels are lower than 0.10.

13.7.2 Analysis and results

13.7.2.1 The characteristics of VFR tourism in Surinam

In Section 13.3, various characteristics of VFR tourism were derived on the basis of a literature review. It is interesting to examine whether Surinamese VFR tourism is in agreement with these characteristics.

This study confirms the earlier findings that VFR tourists are much more likely to stay in non-commercial accommodation and more likely to be repeat-visitors than non-VFR tourists. Furthermore, the outcome confirms the findings of earlier research that VFR tourism is, despite the lower level of total expenditures, economically interesting because of the significant spending on food and beverages and entertainment. But the overall level of spending is lower for VFR tourists because of their lower spending on accommodation. A comparison with other expenditure categories again confirms the finding that VFR tourism is economically interesting. The level of expenses on shopping is the same for VFR tourists as it is for non-VFR tourists (around €280). The expenses on gifts to friends and relatives are significantly higher for VFR tourists than they are for non-VFR tourists (€307 against €137). The same is true for the other expenses of VFR tourists, which are also significantly higher than they are for non-VFR tourists (€33.90 against €11.47).

However, not all characteristics of VFR tourism (found by means of our literature search) are confirmed by our study. The higher package expenses by non-VFR tourists, the longer length of stay for VFR tourists, and the characteristic that VFR tourists are more likely to stay in urban regions are not confirmed by this study.

Besides the above-mentioned differences between VFR and non-VFR tourists, other significant differences were found as well. VFR tourists were younger, lower-educated, more likely to be of Surinamese origin, and more likely to travel alone or with their family or household.[4] Non-VFR tourists were more likely to travel with friends, acquaintances or colleagues. As expected, most VFR tourists had VFR as the main purpose of their visit, followed by leisure. Non-VFR tourists had mostly leisure as the main purpose of their visit, followed by business, study/internship, and other.

13.7.2.2 The heterogeneity within the VFR market in Surinam

The purpose of this study was to determine whether the VFR tourism market is heterogeneous in economic terms. In order to answer this question, four typology

---

[4]   The survey questionnaire did not include the category 'spouse/partner' in the question relating to travel companions. However, it might be assumed spouses and partners are included in the categories 'friends and/or acquaintances' and 'household/family'.

factors have been used in our study: sector; accommodation; migration; and generation. The most important findings on these factors are the following:

- Sector (VFR as main purpose versus VFR as activity)
  - *Socio-demographic and trip-related characteristics*: VFR tourists with VFR as an activity are less likely to be of Surinamese origin and to have been in Surinam before than VFR tourists with VFR as their main purpose. They are higher-educated, less likely to travel alone, and more likely to travel with friends/acquaintances than tourists with VFR as their main purpose.
  - *Level of total expenses*: The total expenses of the two groups did not differ significantly from each other.
  - *The percentage distribution of expenditures*: The percentage distribution of expenditures (see Table 13.4) differed for three out of 18 expenditure categories. For two other categories, indicative differences are found.
- Accommodation used (commercial VFR, that is those staying at least one night in hotels and other commercial accommodation versus non-commercial VFR)
  - *Socio-demographic and trip-related characteristics*: Commercial VFR tourists are less likely to be of Surinamese origin, and are less likely to have been in Surinam before. They are higher-educated and have a higher level of income than VFR tourists who stay only in non-commercial accommodation. Furthermore, commercial VFR tourists are more likely to travel with friends and relatives and more likely to stay outside Paramaribo than non-commercial VFR tourists. Non-commercial VFR tourists are more likely to travel alone, and are more likely to have VFR as the main purpose of their visit. Most commercial VFR tourists also have VFR as the main purpose of their visit, but they are more likely to have leisure or nature as their main purpose of visit than non-commercial VFR tourists.
  - *Level of total expenses*: Commercial VFR tourists had significantly higher total expenses than non-commercial VFR tourists.
  - *The percentage distribution of expenditures*: The percentage distribution of expenditures (see Table 13.4) differed significantly for 10 expenditure categories and indicatively for one category. The finding that commercial VFR tourists are more likely to stay outside Paramaribo than non-commercial VFR tourists is in accordance with their higher proportion of spending on tours. Commercial VFR tourists spent lower proportions of their spending on 'shopping' and 'gifts to friends and relatives' than non-commercial VFR tourists.
- Migration (ethnic VFR tourists versus non-ethnic VFR tourists)
  - *Socio-demographic and trip-related characteristics*: Ethnic migrant VFR tourists are younger, more likely to live in one of the large cities in the Netherlands, more likely to be female and lower-educated than non-ethnic VFR tourists. Furthermore, they are more likely to travel alone or

with children; more likely to have VFR as the main purpose of their visit; more likely to have been in Surinam before; and more likely to stay with friends and relatives than are non-ethnic VFR tourists. The last two findings are not very surprising – after all, ethnic VFR tourists are of Surinamese origin, and it is therefore likely that they have relatives and friends in Surinam. Non-ethnic VFR tourists are more likely to stay in other accommodation than the houses of friends and relatives, and are more likely to stay at least one night outside Paramaribo than are ethnic VFR tourists.

o *Level of total expenses*: Ethnic VFR tourists had significantly higher total expenses than non-ethnic VFR tourists.

o *The percentage distribution of expenditures*: The percentage distribution of expenditures (see Table 13.4) differed significantly for six categories and indicatively for two other categories. The higher proportions of their spending on tours and accommodation by non-ethnic VFR tourists agrees with the finding that they are more likely to stay outside Paramaribo, and are more likely than ethnic VFR tourists to stay in other accommodation besides the houses of friends and relatives.

- Generation (first-generation VFR tourists versus successive generation VFR tourists)

    o *Socio-demographic and trip-related characteristics*: As expected, first-generation VFR tourists are older than the successive-generations VFR tourists. An indicative difference is found for the level of income. First-generation VFR tourists have a higher level of income than successive-generation VFR tourists. An indicative difference is found for only one trip-related characteristic: successive-generation VFR tourists are more likely than first-generation VFR tourists to stay in an apartment/house or in other types of accommodation in Paramaribo.

    o *Level of total expenses*: First-generation VFR tourists had significantly higher total expenses than successive-generation VFR tourists.

    o *The percentage distribution of expenditures*: The percentage distribution of expenditures (see Table 13.4) differed significantly for three categories and indicatively for two other categories. First-generation VFR tourists spent a greater proportion of their spending on 'gifts to friends and relatives': this is probably caused by their stronger family relationships.

Table 13.4 presents the most important outcomes with respect to the economic variables of the different groups of VFR tourists.

Table 13.4 The economic heterogeneity within the VFR market

	VFR	VFR MP	VFR ACT	NCOM VFR	COM VFR	ETH VFR	NETH VFR	FIRST VFR	SUCC VFR
Total expenses[1]	3,167.14	3,202.88 .586	3,056.51	3,054.14 .014**[2]	4,011.11	3,234.09 .036**	2,769.76	3,348.54 .010**	2,710.38

The distribution of the total expenses

	VFR	VFR MP	VFR ACT	NCOM VFR	COM VFR	ETH VFR	NETH VFR	FIRST VFR	SUCC VFR
Package	.31	.16 .207	.78	.13 .352	2.31	.14 .374	1.36	.17 .578	.00
Flight	65.47	66.87 .027**	61.88	66.34 .002**	55.72	65.89 .277	62.85	65.05 .077*	69.85
Accomm.	.96	.90 .579	1.15	.00 .000**	11.84	.53 .000**	3.67	.55 .653	.41
Tours[3]	1.80	1.67 .334	2.18	1.55 .000**	4.69	1.53 .003**	3.48	1.45 .507	1.86
Day trips	.89	.82 .365	1.10	.84 .135	1.54	.84 .313	1.21	.75 .157	1.26
Overnight trips	.91	.85 .570	1.09	.71 .000**	3.15	.68 .002**	2.30	.70 .837	.60
Terraces/ Bar /Café	2.94	2.67 .027**	3.79	2.73 .521	5.08	3.07 .818	2.17	3.09 .864	2.93
Food&Bev &Entertain	7.77	7.25 .027**	9.43	7.87 .521	6.87	7.73 .818	8.02	7.70 .864	7.90
Snack	.86	.88 .777	.81	.89 .420	.56	.89 .510	.68	.96 .252	.59
Rest.	3.36	3.16 .144	3.99	3.39 .835	3.21	3.33 .725	3.58	3.36 .813	3.19
Terr./bar	1.85	1.67 .096*	2.43	1.84 .617	2.15	1.73 .080*	2.61	1.79 .461	1.44
Disco/ Casino	1.58	1.42 .161	2.07	1.63 .030**	.87	1.66 .286	1.05	1.45 .069*	2.64
Theatre /concert	.13	.12 .827	.14	.14 .000**	.00	.14 .007**	.03	.16 .434	.08
Shopping	9.23	9.24 .971	9.19	9.36 .038**	6.79	9.49 .081*	7.59	9.61 .648	8.92
Duty free	.56	.57 .809	.53	.58 .586	.44	.61 .020**	.27	.61 .855	.58
Super-market	3.74	3.52 .213	4.47	3.87 .096*	2.56	3.93 .040**	2.58	4.34 .016**	2.01
Souvenirs	2.15	2.13 .828	2.25	2.12 .434	2.72	2.13 .798	2.29	2.16 .787	2.00

Cloths	2.28	2.49	1.60	2.38	.92	2.33	1.91	2.07	3.58
		.078*		.000**		.628		.297	
Other shop.	.49	.55	.29	.41	.13	.49	.48	.44	.75
		.444		.465		.929		.462	
Gifts to f&r	10.48	10.67	9.87	10.89	6.51	10.50	10.33	11.07	7.82
		.526		.028**		.917		.030**	
Other exp.	1.05	0.80	1.84	1.14	.13	1.13	0.58	1.30	0.30
		.121		.000**		.373		.003**	
Sum	100.01	100.25	100.08	100.03	99.84	100.02	99.95	100.02	100.03

Notes:

1. Respondents were asked to fill in their total expenditures and their expenditures per category. The sum of the expenditures per category appeared to be lower than the total expenditures for all the groups of VFR tourists. Because the total expenses (filled in by the respondents) are more reliable than the sum of the expenditures per category, total expenses are presented in this table. However the percentage distribution of the expenses is based on the expenditures filled in per category by the respondents.
2. * stands for an indicative difference, meaning: significance level is lower than 0.10; ** stands for a significant difference, meaning: significance level is lower than 0.05.
3. The variables tours, F&B&Ent. And Shopping are subtotals. Tours is the sum of day trips and overnight trips (o.n. trips); F&B&Ent. is the sum of snack, restaurants, terraces/bar, disco/casino and theatre/shopping; and shopping is the sum of duty free, supermarket, souvenirs, clothes and other shopping.

13.8 Conclusions and recommendation

This study confirms earlier findings that the VFR tourists are economically interesting because of their considerable spending on food and beverages and retail goods. Moreover, this study has demonstrated the considerable spending on gifts to friends and relatives by VFR tourists.

The study has undoubtedly some limitations in terms of sample size, non-response, gender and age bias, non-exhaustive sample, seasonal bias, reliability of arrival data and actual expenses. Nevertheless, the statistical data provide the best possible and most reliable information on VFR tourism in Surinam. However, the tentative findings in this study seem plausible and are highly interesting.

The presumption that there is heterogeneity within the VFR market is confirmed by this study. Different types of VFR tourists differed from each other not only in terms of socio-demographic and trip-related characteristics, but also in terms of their spending patterns. The distinctions into commercial and non-commercial VFR tourists and ethnic and non-ethnic VFR tourists are useful from an economic perspective. Both the total expenditures and their distribution were influenced by the factors 'accommodation used' and 'migration'. VFR tourists with VFR as the main purpose of their visit, and VFR tourists with VFR as an activity differed only in a few aspects from each other. The same is true for the distinction into first-generation and successive-generation VFR tourists. The expenses on gifts to friends and relatives were significantly higher for the first-generation VFR tourists. A distinction can be useful when these gifts are included in the economic analysis. However, more economic research is necessary in order to determine if the typology of Moscardo et al. (2000) is economically relevant, and if an extension with the typology factor migration is only useful in Surinam or if it is useful in other countries as well. The principal question concerning whether a typology of VFR tourism is economically relevant is therefore hard to answer. But what we can say is that the results of this study and the studies of Lehto et al. (2001), Moscardo et al. (2000) and Seaton and Tagg (1995) indicate that there is heterogeneity within the VFR market in economic terms. This research is the first to compare the spending patterns of ethnic VFR tourists with those of non-ethnic VFR tourists, and is also one of the first to include gifts to friends and relatives as an economic variable. It appears that these gifts form a great part of the expenditures of the VFR tourists. Particularly the first-generation VFR tourists spent a great part (more than 11 per cent) of their total expenditures on gifts. Although the recommendations for tourism statistics advise not to include these gifts in economic tourism analyses, this advice should be reconsidered for the case of VFR tourism. The importance of gifts found by this study is probably explained by the fact that Surinam is a developing country. Therefore, it would be interesting to undertake further research to see whether there are differences in VFR tourism between developing and developed countries.

References

Adams, R.H. and J. Page (2003), *International Migration, Remittances and Poverty in Developing Countries*, World Bank Policy Research Working Paper 3179, World Bank, Washington.

Asiedu, A. (2003), *Some Benefits of Migrants Return Visits to Ghana*, International Workshop on Migration and Poverty in West Africa, Sussex Centre for Migration Research, Sussex (http://www.sussex.ac.uk/migration/research/transrede/workshop/IWMP7.pdf).

Braunlich, C.G. and N. Nadkarni (1995), 'The Importance of the VFR Market to Hotel Industry', *Journal of Tourism Studies*, 6(1): 38–46.

Frechtling, D.C. (1994), 'Assessing the Economic Impacts of Travel and Tourism – Introduction to Travel Economic Impact Estimation', in J.R. Brent Ritchie and Ch. R. Goeldner (eds), *Travel, Tourism and Hospitality Research*, 2nd edn, John Wiley & Sons, New York.

Gamage, A. and B. King (1999), 'Comparing Migrant and Non-migrant Tourism Impacts', *International Journal of Social Economics*, 26(1 2 3): 312–24.

Hu, B. and A.M. Morrison (2002), 'Tripography: Can Destination Use Patterns Enhance Understanding of the VFR Market?', *Journal of Vacation Marketing*, 8(3): 201–20.

Jackson, R.T. (1990), 'VFR Tourism: Is It Underestimated?', *Journal of Tourism Studies*, 1(2): 10–17.

King, B. (1994), 'What Is Ethnic Tourism? An Australian Perspective', *Tourism Management: Research, Policies, Planning*, 15(3): 173–176.

Lehto, X.Y., A.M. Morrison and J.T. O'Leary (2001), 'Does the Visiting Friends and Relatives' Typology Make a Difference? A Study of the International VFR Market to the United States', *Journal of Travel Research*, 40: 201–12.

Liu, J., T. Var and A. Timur (1984), 'Tourist-income Multipliers for Turkey', *Tourism Management: Research, Policies, Planning*, 5(4): 280–287.

McIntosh, R.W., C.R. Goeldner and J.R. Brent Richie (1995), *Tourism, Principles, Practices, Philosophies*, Wiley, New York.

Meis, S., S. Joyal and A. Trites (1995), 'The U.S. Repeat and VFR Visitor to Canada: Come Again, Eh!', *Journal of Tourism Studies*, 6(1): 27–37.

Morrison, A.M., S. Hsieh and J.T. O'Leary (1995), 'Segmenting the Visiting Friends and Relatives Market by Holiday Activity Participation', *Journal of Tourism Studies*, 6(1): 48–62.

Morrison, A.M. and J.T. O'Leary (1995), 'The VFR Market: Desperately Seeking Respect', *Journal of Tourism Studies*, 6(1): 2–5.

Moscardo, G., P. Pearce, A. Morrison, D. Green and J.T. O'Leary (2000), 'Developing a Typology for Understanding Visiting Friends and Relatives Markets', *Journal of Travel Research*, 38: 251–9.

Müri, F. and A. Sägesser (2003), 'Is VFR an Independent Target Group? The Case of Switzerland', *Tourism Review*, 58(4): 28–33.

Noordewier, T. (2001), *2001 National Survey of the Vermont Visitor: An Examination of the Friends and Relatives (VFR) Traveller*, University of Vermont, Vermont (http://www.uvm.edu/~snrvtdc/publications/visiting_friends_relatives.pdf).

OECD (1997), *Tourism Policy and International Tourism in OECD Countries*, OECD, Paris (http://www.oecd.org/dataoecd/31/8/2755255.pdf).

Osaki, K. (2003), 'Migrant Remittances in Thailand: Economic Necessity or Social Norm?', *Journal of Population Research* (November) (http://articles.findarticles.com/p/articles/mi_m0PCG/is_2_20/ai_111014984).

Ostrowski, S. (1991), 'Ethnic Tourism, Focus on Poland', *Tourism Management: Research, Policies, Planning*, 12(2): 125–30.

Seaton, A.V. and C. Palmer (1997), 'Understanding VFR Tourism Behaviour: The First Five Years of the United Kingdom Tourism Survey', *Tourism Management: Research, Policies, Planning*, 18(6): 345–55.

Seaton, A.V. and S. Tagg (1995), 'Disaggregating Friends and Relatives in VFR Tourism Research: The Northern Ireland Evidence 1991–1993', *Journal of Tourism Studies*, 6(1): 6–18.

Stynes, D.J. (2001), *Economic Significance of Tourism to the Greater Lansing Economy* (http://www.prr.msu.edu/miteim/satellite/Lansing97sat.pdf).

Theobald, W.F. (ed.) (2004), *Global Tourism*, Butterworth-Heinemann, Oxford.

Theuns, H.L. (2002), 'Tourism and Development: Economic Dimensions', *Tourism Recreation Research*, 27(1): 69–81.

United Nations and WTO (2000), *Recommendations on Tourism Statistics* (Rev-1.0), United Nation, New York (http://unstats.un.org/unsd/statcom/doc00/m83rev1.pdf).

Williams, A.M. and C.M. Hall (2000), 'Tourism and Migration: New Relationships between Production and Consumption', *Tourism Geographies*, 2(1): 5–27.

Wood, R.E. (1998), 'Touristic Ethnicity: A Brief Itinerary,' *Ethnic and Racial Studies*, 21(March): 218–41.

WTO (2001), *Tourism Market Trends: The World*, WTO, Madrid.

Yuan, T., J.D. Fridgen, S. Hsieh and J.T. O'Leary (1995), 'Visiting Friends and Relatives Travel Market: The Dutch Case', *Journal of Tourism Studies*, 6(1): 19–26.

Chapter 14

Technology Transfer to Small Businesses in the Tourist Sector: Supporting Regional Economic Development

Peter M. Townroe

14.1 Introduction

In many areas of modern economies, advances in communications technologies have combined with emerging management philosophies to foster and encourage successful smaller and medium-sized enterprises to exist alongside larger corporations. Two forces can be seen to have been at work in this respect over the past two decades or so.

The first such force has been a retreat by many larger companies from the attempt to hold together a very diversified portfolio of activities in a wide span of subsidiary companies. By the 1960s, computer-based accounting systems and easier worldwide telephone and data processing links had permitted a degree of informed central control by a head office of large numbers of subsidiary companies active in different areas of industry and based in different wordwide locations. While these advantages remain, and indeed have been enhanced over the past 30 years, many managements came to realize that the advantages had more to do with gains in static efficiency than with gaining dynamic advantage over competitors. Dynamic advantage, the key to long-run success, has much more to do with the generation of new products and with pushing forward technologies of production.

The strategic concentration by companies on sustaining and building upon their dynamic advantage leads to the realization that maintaining managerial thrust in many activity areas simultaneously is a major problematic; and that many corporate activities that are subsidiary to the delivery of the main product or products can best be purchased in from specialist providers rather than organized as component parts of the home organization. This will be the case even when many of these specialist providers of support activities are themselves quite small enterprises.

Alongside this change in management philosophy, and the refining of subcontract relationships with these specialist providers, has come the second force in favour of smaller companies. This has involved the power of modern communications technologies to provide smaller enterprises with both the internal control mechanisms and the external supplier and market information which have been required in order to compete with larger players in the market place. These changes have been much discussed in the management and industrial economics literatures in relation to the manufacturing sectors of economies. There has been some discussion in areas like transportation, property, retailing and the utilities (see, for example, Storey, 1994); but very little in relation to the tourist and leisure sector.

An awareness of the changing balance within economies between large enterprises and smaller companies has been an important driving force behind the slew of government policies in the past two decades to assist and enhance the competitiveness of the smaller companies. Governments have realized that growth in national output per capita requires rising rates of labour productivity from these smaller companies, as much as from the larger enterprises with which the smaller firms compete or to which they provide inputs. And this rising productivity from the smaller companies puts competitive pressure on their larger brethren and/or reduces the costs of commodity or service inputs. Either way, international competitiveness in both import and export markets is enhanced. Within the European Union this has been a major policy focus for Member States and for the Community as a whole for the past 40 years, but especially under the Single Market measures of 1992.

The concern of this chapter is with one small corner of this policy attention by governments of the Member States of the European Union for the vibrancy of the small and medium-sized enterprise (SME) sector of their economies. However, it is a small corner that is particularly important for a number of the regions and sub-regions within the Union. The tourist industry in Europe has large corporate players: hotel chains, airlines, holiday tour companies, major theme-park operators, fast-food chains, car rental companies, providers of key infrastructure services, etc. But it is also an industry of many small players: family-sized businesses, individual entrepreneurs starting out for the first time, franchisees, small firms operating from a single location, etc. These tourist industry SMEs are all users of technology, in many different ways. This chapter is about how these smaller enterprises might more readily gain from the technologies that are available to them; or gain assistance in order to generate their own new technologies.

Across the Member States of the European Union, there is now a multitude of public and semi-public agencies charged with the task of supporting the SMEs based in their local area (see, for example, Bennett and McCoshan, 1993). Frequently, these agencies have been given budgets to provide elements of subsidy and financial support to small enterprises deemed to be deserving on one or another criteria. They have an information dissemination role, pushing out facts and figures to busy entrepreneurs. And in many cases they have a training and education role, for both the entrepreneur and his or her staff. Within this role, some of these agencies have a responsibility to give support to the acquisition and development of technology. Elsewhere, public sector support for technological development is channelled through different agencies or institutes. This chapter, written in 1997, will use the illustration of one such channel to suggest ways of assisting tourist-related businesses in this respect, especially those businesses located in lagging regions or sub-regions.

14.2 Tourist-related activity

'Tourist-related activity' is difficult to define in terms of a standard statistical classification. Activities which are related to the expenditures by tourists overlap into the forms of expenditure which arise from many other social groups or from companies. Business people use hotels, local residents use bars and restaurants, local school children visit museums, and all categories use modes of transport. And 'tourist-related' is frequently linked to 'leisure activities'. And 'leisure activities' may be classified into 'within the home' and 'outside the home'. Participant and observer sport is a major area of expenditure of the second category for example, but not usually perceived as 'tourism'.

Many areas seeking to build up their local tourist industry will see the demand from local residents and from day-trip visitors as being as important as the demand from overnight stays. In economic terms, the day-trip visitors from outside the area combine with the overnight stay visitors from homes also outside the local area, and with international overnight stay visitors to contribute to the revenues of the economic base of the area; and through these revenues to contribute to local economic output and to employment.

In 1992, Eurostat listed the following economic sectors as 'tourist-related activities':

- hotels, guest houses or similar;
- restaurants and bars;
- coffee bars and public houses;
- travel agencies;
- car rentals;
- libraries, archives and museums;
- public tourist boards and offices;
- pleasure ports;
- others.

(From the published report it is not clear what the meaning of 'pleasure ports', or 'touristische hafen' or 'ports de plaissance' might be, but the only Member State to record having these was Denmark, with 314 in 1992.)

The same report records the United Kingdom as having 1.68 million employees involved in tourist-related activity. This is nearly 6 per cent of the labour force. In 1994/95, there were 21.5 million visitors to the United Kingdom from other countries, spending an estimated £10.3 billion. In the same year, there were 40.1 million visits out of the country by UK residents, spending an estimated £14.5 billion. International tourism is the third largest foreign exchange earner in the British economy. The UK ranks fifth in the world in earnings from tourism. (In 1992 the UK received 128,000 visitors from Greece, but sent 2,154,000 visitors).

At a sub-national scale two British examples can give an impression of the importance now being attached to the local and sub-regional economic contribution of tourist-related activity.

The first of these is the City of Manchester. 'Urban tourism' has come to be an important strand of local economic development activity in many towns and cities across Europe; and not just in capital cities and in towns noted for their particular histories. Towns and cities which have grown up on industrial activity in the past 200 years have also been discovering that they can put themselves forward as tourist destinations, and in many cases have been investing in local facilities and attractions to increase their market share. It is now all but mandatory that policies of urban economic and social regeneration do not fail to consider a tourist activity dimension (for example, Law, 1993).

The 1993 Census of Employment found that, in the area covered by the Manchester Training and Enterprise Council, an area with a labour force of 432,000 which covers the four core local authorities of the Greater Manchester area (i.e. the City of Manchester, Salford, Tameside, and Trafford), some 30,600 people could be classified as employed in the recreation, tourism and leisure sector of the local economy. They were working in 2,600 firms, with a mean size of 12 employees. A further 1,100 people were self-employed in the sector. Of the employees, 43 per cent were female and part-time, 15 per cent were male and part-time, and 17 per cent female and full-time, and 25 per cent male and full-time.

Manchester hosted the 2002 Commonwealth Games, a major tourist event but also an opportunity to invest in new facilities for use thereafter. These included a new £90m 'Millenium Stadium'. Other key investments which now support the development of the local tourist market have included the new Bridgewater concert hall, the extensions to the City art museum and the Lowry Centre, all supported by Lottery money. The Lowry Centre was planned to attract 750,000 visitors per year, the third largest attraction in North West England. The Granada Television Castlefield Studio Tours complex has been expanded. The G-Mex Exhibition Hall has been complemented with a large investment in retail and convention facilities. And Manchester Airport now has a second runway, allowing the annual throughput of passengers to exceed London Gatwick and rise to 30 million plus per annum. All of these investments illustrate what has happened in many cities seeking to diversify their economic base. (For a fuller discussion of the Manchester case, see English Tourist Board, 1989).

The second example is provided by the county of Hampshire in the south of England. Hampshire is a diverse county of 1.62 million residents, stretching from an area of London commuter suburbs and of hi-tech aerospace and defence-related industry in the north of the county to the civilian and naval ports of the

Southampton–Portsmouth agglomeration in the south. The majority of the land area of the county is rural, with the cathedral county town of Winchester in the centre. By national standards, the county is prosperous, with relatively high earnings and low rates of unemployment.

The Hampshire Training and Enterprise Council (Hampshire TEC, 1997) has nominated 'Leisure and Tourism' as one of six key sectors in the local economy, along with Aerospace, Information Technology, Marine Industries, Construction, and Financial Services. Tourism in the county is estimated to employ about 50,000 of the labour force of 750,000. There are about 3 million trips by staying visitors each year, with £380 million of associated expenditure. This is complemented by a further £100 million from day-trip visitors. The diversity of the tourism attractions in the county (which include cultural and historic heritage sites, protected open countryside in the New Forest, water-based activity centres along the coast, naval exhibits in Portsmouth, and many attractive inland villages) matches well with the growth markets in tourism.

At the same time, employment in the sector is over 50 per cent part-time and female, and poorly paid. Many jobs are seasonal and involve unsocial hours. Employers in the sector have a poor image. Recruitment is a problem, even though the required skill levels for many openings are low with little need for formal qualifications. Experience in Hampshire accords with the findings of a 1995 survey undertaken by the Confederation of British Industry into the skill gaps in the tourist industry as shown in Table 14.1.

Table 14.1 Tourist industry skill gaps

Skill gaps	Percentage of businesses ranking the named gap as most important
Craft or practical skills (incl. catering)	24
Customer care skills	12
Financial awareness	12
Personal qualities	10
Languages	10
IT	7
People management	5

Neither 'Management and Administration' nor 'Marketing and Sales' were ranked as most important, a finding that is out of line with the findings of other surveys of SMEs which point to both of these areas as being typically deficient in many smaller firms (for example, Barber et al., 1989). Perhaps, as the Hampshire TEC report suggests, tourism is an industry that improves its competitiveness by employing workers with a combination of technical, commercial, creative and communication skills.

In regions and sub-regions characterized by relatively low incomes and by high levels of un- or underemployment, the pursuit of economic development through a tourism development policy has two attractions (Townroe and Mallalieu, 1993). The first is the low levels of capital investment typically needed to support the start-up of new businesses or the expansion of existing businesses. The risk levels for commercial lending may be high, but public sector subsidy can be justified on two grounds: that entrepreneurial learning takes place even in a business failure, and that fixed capital (for example, of a hotel bedroom or of a restaurant kitchen) remains locally for the next owner. And the second attraction is that the small scales of operation in many tourist-related activities provide a good training ground for entrepreneurs who may then progress to build up larger enterprises not only in tourism but also in other industries within the region (Williams and Shaw, 1988).

14.3 Technology in tourism

Tourist-related activities are not normally linked to generalized discussions of 'Technology'. Tourist activities are seen as low-tech. The perception is of an industrial sector of low capital intensity, with little regard for the processes of invention and innovation: an industry where research and development activity is all but unknown, where the technology is handed down from other sectors. While there is some truth in this characterization, for without some well-founded basis the view would not persist, a moment's reflection will lead to the realization that in fact this is an industrial sector that is an intensive user of many technologies, and that in its own way it is also a generator of many technologies, even if not so many of them are hi-tech in the conventional sense.

The technologies being used in the tourist industry are both 'hard' and 'soft'. The hard technologies mostly come from other industries, but frequently to a design specification drawn up within the tourist industry. The construction industry builds hotels to designs by architects working to a close brief; the design and construction of theme park rides is a specialist sub-sector of the mechanical engineering industry; the fast food business is dependent on semi-prepared food items; the commercial kitchen equipment industry is different to the industry which supplies households with their kitchen requirements; libraries, museums and archives each have their range of specialized items of hardware. In many of these instances, the momentum for technological advance comes from individuals within the tourist and leisure industry businesses, rather than from the equipment supplier. This will be especially true where one-off designs are involved and/or new designs are important elements in competitive advantage (as in theme park rides for example).

However, most technologies in the tourist industry that are both developed in and used in the industry are 'soft' technologies: ways of doing things, ways of using hardware, ways of putting together a tourist experience. These will be of two kinds: product designs to gain market advantage, and cost-cutting and efficiency-enhancing changes in both the product itself and in the process of delivering the product to the customer.

Many product technology changes are to do with design: of buildings, of rooms, of decor, of literature, and of equipment where that is important, as in theme parks, fun-fairs, presentational experiences, etc. (There is, of course, a huge souvenir industry linked to tourism, backed into other industries). The cost-cutting technologies are also to do with design and with servicing as well as with management, of staff as well as of the customers. There are also technologies which arise from needing to respond to various legislations, for safety in particular. These include fire regulations, mechanical safety and the safety of food and drink. There are also land-use planning and environmental regulations which have particular relevance for certain types of tourist-related investments. In all of these areas innovations arise from within the industry.

Bringing the focus to a given region or sub-region, advances in technologies, both hard and soft, of relevance to the tourist industry may be brought to local players through six different 'pathways' (Townroe, 1990).

The first of these is by far and away the most common in the tourist industry, as it is indeed in most other industries. This is the 'off-the-shelf' purchasing from suppliers of new plant and equipment which embodies new capabilities. For small businesses, the principal issue here is being aware of what is available on the relatively infrequent occasions when purchases are made. The second is the pathway that comes with migrant businesses, locating a new branch in the region or transferring operations from elsewhere and bringing technologies with them. The third pathway is through the intra-company transfer of skills and know-how between subsidiary units and into the region in question. In the tourist sector, the mechanism of the franchise is important in distributing new expertise in this way.

The fourth pathway can be more informal in terms of how a new technology is developed. This is the generation of a technological advance from within an existing company. The spur for this may be reactions from customers, or it may be from ideas that arise in marketing or in product delivery. There may be some research and development expenditure involved but it will rarely be called that.

The fifth pathway is a form of externality from one regional business to another. It is brought about by local flows of information and by demonstration. One business learns from another. The learning company needs to have the ability to understand and assimilate and then apply the new knowledge. There are agglomerations economies in this respect in the tourist industry, the local milieu can be important, as suggested by Camagni (1991) and in the sense of Davelaar (1991), in support of innovative activity (see also Davelaar and Nijkamp, 1997).

The sixth pathway relates to this idea of the local availability of knowledge. This is the use of local experts or consultants to respond to a problem or to identify an opportunity, with a new piece of hardware or with a new way of doing something. This pathway may well involve a local college or university.

All six of these pathways channel innovations which, if successful, will raise the productivity of capital and labour in the industry in the region. And in doing so, they may attract new capital and labour into the region at the margin. However, there are skills required to take advantage of innovations as well as to generate them. These will be knowledge-based, diagnostic, flexible, and, above all in the tourist industry, entrepreneurial. The regional economic development challenge in many parts of the tourist industry comes from facing players who are unresponsive to new ideas and who are resistant to change.

14.4 A technology support system

Can public policy intervene effectively to provide support for technological advance in the tourist industry at the local or sub-regional level? Mechanisms of business support already exist in many regions and sub-regions of the EU, as noted above. Guidance is available in areas such as marketing, cash flow management, labour law, project appraisal, customer care, etc. One possible route to providing assistance to businesses in the tourist industry on matters technological is clearly to work through the existing business support agencies: chambers of commerce, regional development agencies, local government economic development units, etc.

Three problems exist with this approach. The first is a lack of focus. Many entrepreneurs in smaller businesses in the tourist industry in particular will find it hard to identify with managers from larger enterprises in other branches of industry in the region. They do not see the concerns of these managers as being their concerns. Even the nature of the profit objective and the reason for being in business will be different for many. And the very notions of invention and innovation are framed and understood differently (Sweeney, 1987). Therefore the all-encompassing business support agencies may well seem remote from the problems and the ambitions of SMEs in the tourist sector. And the agencies themselves are unlikely to have staff with specialist knowledge of the tourist sector.

The second problem links with the first and that is a lack of ownership. Business support agencies have to work through local companies and that requires commitment by the companies of both money and time; and for most small businesses time is the most important constraint in investing in growth and expansion. Surveys tell us that many small business entrepreneurs tend to be very reluctant to turn to anyone for advice on anything. The 'go-it-alone syndrome' is strong (Blackburn and Curran, 1993). A business support initiative within a local area, if it is to do more than just pay out financial subsidies, requires a sense of ownership by the prospective clientele. There has to be some sense of sharing, even among local competitors.

The third problem in directing technological assistance to the tourist sector in a region through a general business support agency lies in gaining the enthusiasm, commitment and support of institutions in the region that are able to provide the technological inputs. These institutions will normally be local universities and colleges, but may include other private or public sector research organizations.

Two recent technology transfer initiatives in the Yorkshire and Humberside region of the United Kingdom may provide elements of a blue-print for a regional technology transfer policy for the tourist industry in a region, as part of an overall regional economic development policy.

14.4.1 The Food Technopole

The Yorkshire and Humberside Food Technopole is an initiative from 1996 to support the Food, Drinks and Catering industry within those parts of the Yorkshire and Humberside region that are classified as Objective 2 and Objective 5b areas. The initiative is part-funded by the European Regional Development Fund (ERDF) and assistance is therefore restricted to SMEs based within the designated areas (the SMEs being defined as businesses with less than 250 employees or with a turnover of less than £16 million).

The Technopole is a grouping of partners from within the region with expertise in food and drinks manufacture and research and with expertise in technical training and applied business advice. The partners are from three different types of organization:

1. academic and research partners;
2. business support and economic development partner;
3. food industry bodies.

The Technopole is managed by The Food Innovation Centre, an organization that is jointly funded by the University of Lincolnshire and Humberside and the Ministry of Agriculture, Fisheries and Food. It is based in Grimsby.

The Technopole offers assistance to companies in the industry in a number of different ways. Available at zero cost to the companies is one half-day of consultancy, a Training Needs Analysis, and information disseminated directly or via the Internet or by sign-posting companies to relevant individuals within the partner organizations. The free consultancy may be very focussed on a previously identified problem; or may involve a wider review of a sub-set of operations within the company, leading to recommendations for action or for further study.

The Technopole offers two services for which the recipient company is expected to make a financial contribution. The first is what is termed an 'Opportunity Audit'. This involves an audit of the products and the technologies of a company, with an assessment of what the business might be able to do which it is not presently doing. This is followed by a 'controlled brainstorm', involving the auditor, a representative of the firm, a Personal Business Advisor, an academic, and a facilitator. Opportunities for the company which are prospectively commercially viable are generated, with a list of actions beside them. An Action Plan is subsequently drawn up leading to a feasibility study. The ERDF pays for half the cost involved, which usually comes to between £1,200 and £2,000. Between 3 and 5 days of consultancy time is involved.

The second service offered by the Technopole to which a recipient company has to contribute financially, as a sign of commitment, is Project Development. The expenditure here is on consultancy assistance, typically from one of the partner universities. A two-thirds subsidy is offered for projects of value up to £15,000. The project may well involve a training dimension.

Types of project include:

- cost saving – increasing efficiency and productivity, waste minimization, energy saving, improvements in food hygiene and quality;
- new product development – innovation, assessment of concept feasibility, solution to problems encountered;
- market penetration – product strengths, access to market information;
- environmental improvements – minimization of raw material usage, reduction of the impact of waste products, utilization of waste and by-products;
- access to appropriate technology – pilot plant and test equipment, demonstrations, evaluation of needs.

The Technopole also offers a range of seminars, conferences and training events at different locations within its region.

14.4.2 The Materials Technopole

The industrial focus of the Sheffield Regional Technopole was on metals and materials manufacturing and on metals-related engineering. Geographically, it covered the Objective 2 area of South Yorkshire in the Yorkshire and Humberside Region and North Derbyshire and North Nottinghamshire in the East Midlands Region. Like the Food Technopole, it was part-funded by the ERDF, at £3.04 million over three years to the end of June 1998. Like the Food Technopole, the focus was on SMEs within its industrial sector.

Also like the Food Technopole, the Sheffield Regional Technopole was essentially a brokering facility. It brought small and medium-sized companies which had a technological problem or a product or a process proposition together with partner organizations of the Technopole which had a technological development capacity. Although the Technopole tried to 'sign-post' inquiries to organizations outside of the Sheffield area if that seemed necessary, the partner organizations of the Technopole were the principal providers of assistance.

The partners are listed in Appendix 14.I. They included the two universities in Sheffield, a number of specialized public and private sector research laboratories, business support agencies, two chambers of commerce, and two local authorities. The lead partner was The Company of Cutlers in Hallamshire, the long-established association of industrial leaders in Sheffield. The Cutlers Company was responsible for the administration of the Technopole and for liaison with the

Government Office for the Yorkshire and Humberside Region and the Commission of the European Union.

The partners could support SMEs in the development of new products and processes in the following areas of technology: steel and steel-based products, springs, castings, precious and base metals, health and safety requirements, glass products, cutlery and cutting tools, casting in metals, materials manufacturing processes, and biological, chemical, environmental, structural, and mechanical analysis and testing. This was achieved with a Voucher Scheme which provided a grant of up to £7,500 towards half the cost of each project. Support was also given through a Bursary Scheme, of up to £2,625 for individual employees to develop their technological expertise, again at half the cost involved.

The Voucher Scheme has been a considerable success. Between 1 July 1995 and 31 March 1997, some 128 technology transfer projects had been set in train under the Voucher Scheme, with a cost to the Technopole of approximately £662,000 of grants awarded, generating £1,851,000 of total project value. The average value of each grant has been under £6,000. The Bursary Scheme has been less successful so far, small companies being reluctant to put up their share of the training cost involved.

The Materials Technopole also had a budget line to finance publicity and promotion of technology in the materials and metals engineering industries. This was used to sponsor seminars, exhibitions and conferences as well as a newsletter and a web site and an Information Centre for Metals and Materials Manufacture based within Sheffield City Council.

14.5 Conclusion

There are a number of lessons to be learned from the (relatively short) experience of the two Technopoles in the Yorkshire and Humberside region. Although both are directed at technological inputs for manufacturing activity in the two industrial sectors concerned, their experience can be brought to bear on the technologies involved in the service activities that surround manufacturing in a company; and on the technologies involved in a wholly service sector-oriented industry, as is the case with the tourist industry.

The first lesson is that formal mechanisms need to be established to make the technological expertise that resides within local universities and colleges and research institutes available to smaller enterprises in a region. Informal inquiry and contact tends to achieve little.

The second lesson is that smaller local companies tend not to seek assistance in the development of new products and processes. Such assistance has therefore to be actively marketed to the relevant community of smaller firms within the region.

The third lesson is that bringing prospective suppliers of technological assistance into a partnership is not easy. There is competition involved, and providing this sort of assistance is not a high priority for these organizations.

Enthusiasm has to be encouraged among the individuals who will be in the front line of advice and support to client companies as well as among leaders and senior managers.

Fourth, supply partners will worry about the full costing of their contributions, universities and colleges in particular sometimes not being very proficient at this. And the partners will rightly worry about the ownership of any intellectual property rights that are generated by the support activity, from entitlement to publish to the registering of property rights.

Fifth, networking with other small business support and advice agencies and with local company representational organisations is extremely important.

And sixth, the support of government (or EU) funds has proved a necessary condition for launching a technology-brokering initiative; and on-going funding support will probably be a condition necessary to maintain on-going success.

These lessons can be applied to a tourist industry focussed initiative. Many universities now offer programmes in food science, hotel and catering management, tourist management studies, heritage management, and travel industry studies. A technology transfer scheme built on this sort of base can provide a new dimension to the support given to the tourist industry in a region under a regional economic development programme. This is what may be termed 'practical technology': small-scale projects with a clear end product in view, meeting a market need or securing a market advantage. But, to be successful, the community of owners and entrepreneurs within the tourist industry in the region must be brought in to give voluble support for the scheme.

References

Barber, J., J.S. Metcalfe and M. Porteous (1989), *Barriers to Growth in Small Firms*, Routledge, London.

Bennett, R.J. and A. McCoshan (1993), *Enterprise and Human Resource Development: Local Capacity Building*, Paul Chapman Publishing, London.

Blackburn, R. and J. Curran (1993), 'In Search of Spatial Differences: Evidence from a Study of Small Service Sector Enterprises', in J. Curran and D. Storey (eds), *Small Firms in Urban and Rural Locations*, Routledge, London.

Camagni, R. (1991), 'Local "Milieu", Uncertainty and Innovation Networks: Towards a New Dynamic Theory of Economic Space', in R. Camagni (ed.), *Innovation Networks: Spatial Perspectives*, Belhaven Press, London and New York.

Davelaar, E.J. (1991), *Regional Economic Analysis of Innovation and Incubation*, Avebury Press, Aldershot.

Davelaar, E.J. and P. Nijkamp (1997), 'Spatial Dispersion of Technological Innovation: A Review', in C.S. Bertuglia, S. Lombardo and P. Nijkamp (eds), *Innovative Behaviour in Space and Time*, Springer-Verlag, Berlin.

English Tourist Board (1989), *Manchester, Salford, Trafford: Strategic Development Initiative: A Framework for Tourist Development*, London.

Hampshire Training and Enterprise Council (1997), *Local Economic Assessment 1996/97*, Fareham.

Law, C.M. (1993), *Urban Tourism: Attracting Visitors to Large Cities*, Mansell, London.

Storey, D.J. (1994), *Understanding the Small Business Sector*, Routledge, London.

Sweeney, G.P. (1987), *Innovation, Entrepreneurs and Regional Development*, St. Martin's Press, New York.

Townroe, P.M. (1990), 'Regional Development Potentials and Innovation Capacities', in H.J. Ewers and J. Allesch (eds), *Innovation and Regional Development: Strategies, Instruments and Policy Coordination*, de Gruyter, Berlin.

Townroe, P.M. and K. Mallalieu (1993), 'Founding a New Business in the Countryside', in J. Curran and D. Storey (eds), *Small Firms in Urban and Rural Locations*, Routledge, London.

Williams, A.M. and G. Shaw (1988), *Tourism and Economic Development*, Frances Pinter, London.

Appendix 14.I Partners in the Sheffield Regional Technopole

Assay Office, Sheffield
British Glass Laboratories
British Steel: Swinden Technology Centre, Rotherham
Business Link, Sheffield
The Castings Development Centre
The Company of Cutlers in Hallamshire
The Cutlery and Allied Trades Research Association
Health and Safety Laboratory
Rotherham Training and Enterprise Council
Rotherham Chamber of Industry and Commerce
Sheffield City Council
Sheffield Development Corporation (to March 1997)
Sheffield Hallam University
Sheffield Chamber of Industry and Commerce
Sheffield Science and Technology Parks
Sheffield Training and Enterprise Council
Spring Research and Manufacturers' Association
University of Sheffield
Yorkshire Cable Communications

Chapter 15

Possibilities for Using ICT to Realize Heritage and Ecotourism Potential: A Case Study on the Zululand Region of South Africa

Peter Robinson

15.1 Introduction

Zululand is a remote rural region with a large population and high levels of poverty and unemployment. During the past ten years its economic base has been adversely affected by the decline of the coal mining industry. More recently, the economy of one of its main towns, Ulundi, which was the joint provincial capital, has been threatened by the loss of this status and the possible removal of a significant number of associated functions, jobs and spending power. Regional and local planning studies have identified heritage and ecotourism potential as an opportunity to diversify the local economy. Much of this tourist potential is vested in the town of Ulundi and in the adjacent eMakhosini valley.

Notwithstanding these opportunities, progress in translating potential into viable tourism products and marketing these has been slow. This chapter will set the area in its regional context, analyze prevailing socio-economic conditions and outline the development challenges being faced. This will be followed by a discussion of the significant potential that exists for heritage and eco-tourism and the obstacles being encountered in attempting to bridge the gaps between tourism potential and deliverable tourism products which can make a contribution to the local economy. Finally, the chapter discusses the limited extent to which ICT is used at present and reflect on whether ICT could be a missing ingredient in the attempt to establish tourism as a pillar of the area's economy.

15.2 The Ulundi and eMakhosini-Ophathe area: Socio-economic conditions

Zululand is located in the northern part of the province of KwaZulu-Natal. It is relatively remote in the sense that it is about 250km from the economic hub of Durban. However, it is bounded by a national road which connects Durban with the inland centres around Johannesburg, while another main route passes through Zululand from the port city of Richards Bay to the interior.

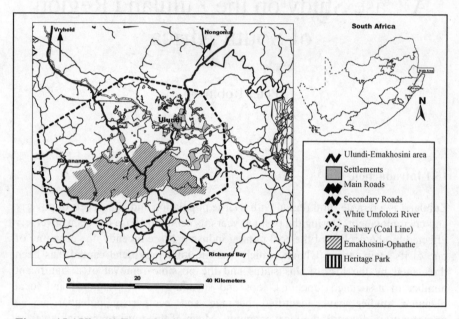

Figure 15.1 Ulundi-Emakhosini area

The historical core of Zululand, which embraces the eMakhosini valley and Ulundi, contains an impressive array of heritage sites relating to settlement and ways of life of the Zulu people, to wars between rival Zulu clans in the early nineteenth century, wars between the Zulus and the Boers, and between the Zulus and the British in the late nineteenth century. The area also contains the Ophathe game reserve. Further afield (some 100km) are other well-known Battlefields such as Rorke's Drift and Isandlwana (Zulu-British wars), and Spionkop (Anglo-Boer wars).

The research on which this chapter is based was defined as an area of some 2,000sq km surrounding the town of Ulundi and the eMakhosini valley. Statistical data was derived from a Water Services Development Plan (WSDP) in 2001, which is regarded locally as the most reliable source of information about the settlement patterns and access to basic services.

15.2.1 Demographic characteristics

Today, the area has a population of 198,000, of whom a large proportion (43 per cent) are under the age of 16. Only 4 per cent are over 65 years of age. This points to a high level of dependency as well as to the wide gap between the number of people entering the labour force in relation to the few job opportunities available. The gender ratio of 54 per cent female to 46 per cent male reflects the absence of men working (or seeking work) in the cities. The local labour force has a similar imbalance. A total of 60,900 children (or 31 per cent) are of school age, almost equally divided between boys and girls. All the people in this area are African and have Zulu as their home language. The population is growing slowly, due to the increasing impact of HIV/AIDS, and is expected to increase by some 32,000 over the next 20 years.

Rural households are typically large, averaging 8.7 persons/household. Table 15.1 shows that, while 70 per cent have eight or more persons, there is a significant number of settlements with household sizes of five or less.

Table 15.1 Size of rural households

Persons per household	No. of settlements	%
Up to 5	18	17
6–7	14	13
8–9	27	25
10–11	30	28
12–13	12	11
14+	7	6
Total	108	100

Source: WSDP (2001).

15.2.2 Settlement pattern

The area has only two towns. The main town, Ulundi, has a population of around 55,000 people, most of whom live in formal dwellings with access to basic services such as water, sanitation, electricity, roads, telephones, schools, clinics, shops and so on. Ulundi's economic base is largely administrative, on account of its establishment in the 1970s as the capital of the KwaZulu homeland. After the democratic transformation in South Africa in the early 1990s, Ulundi was designated as joint capital of the new province of KwaZulu-Natal. For these reasons, a relatively large number of government functions, together with their offices and staff have been located in Ulundi. In addition, it is the seat of the district municipality. However, Ulundi's status as joint provincial capital has been under threat for some years on account of the weakening of the political alliance between the two main political parties in the province. During 2002, this alliance broke down resulting in a decision that there will be a single capital, namely Pietermaritzburg. Some departments and their staff subsequently left Ulundi: Ulundi's economic base and the livelihoods of the thousands of local residents, who derived a living directly or indirectly from the provincial government activities, were under threat. The town also serves as a commercial and service centre for a large rural population.

Babanango, the other town in the area, is no more than a hamlet of some 3,000 people who have access to similar facilities as those living in Ulundi. It was developed as a service centre for the surrounding farming community, but it experienced a period of decline after much of the previously mixed type farming was replaced by large-scale timber plantations, which required much less labour.

The remaining 71 per cent of the population live in 108 rural settlements, which are located on Traditional Authority land. These are scattered and, with a few exceptions, are very small. Table 15.2 shows the distribution of rural settlement sizes. Only 3 per cent of settlements have more than 500 households, while 61 per cent have less than 100 households. These factors, coupled with the broken topography, make servicing difficult and costly. Most of the settlements (82 per cent) have been in existence for 10 to 20 years and 16 per cent for longer. As a result this settlement pattern may be regarded as well-established.

Table 15.2 Settlement sizes

Settlement size (no. of households)	No. of settlements	%
Up to 50	38	35
51–100	28	26
101–150	15	14
151–250	10	9
251–500	14	13
501–2100	3	3
Total	108	100

Source: WSDP (2001).

15.2.3 Access to basic services

In the rural settlements, access to basic services and infrastructure is one of the main development challenges. Analysis of the availability of basic services in the 108 rural settlements revealed that 74 per cent have access to some form of clean water, but in 26 per cent water has to be collected from a spring or river (Table 15.3). Although 31 per cent of settlements have yard connections, almost 90 per cent of the population have to walk more than 1 km to reach their source of water and 48 per cent walk more than 2 km. This work is undertaken almost exclusively by women and children.

Table 15.3 Access to water supply

Type of water supply	No. of settlements	%
Yard connection	27	31
Borehole	37	43
Spring	15	18
River	7	8
Total	86	100

Source: WSDP (2001).
Note: No data for some settlements.

Levels of sanitation are often linked to water supply. By far the vast majority of settlements use pit latrines for sanitation. Only one has a waterborne sewage disposal system. Almost 60 per cent of settlements have access to electricity, but this does not necessarily mean that all households in the villages served will be users. Those without electricity use candles and paraffin for lighting, and a combination of open fires, paraffin and gas for cooking. All these settlements have gravel access roads, most of which are maintained by the provincial or district authorities.

However, it is the combination of basic services that defines the relative welfare of the settlements. Table 15.4 classifies the settlements into those relatively better off (with a combination of clean water – yard or borehole, adequate sanitation, electricity and regularly maintained road access), those with an intermediate level of services, and those that are worse off (usually without electricity and/or inadequate water supply at 2km or more from the settlement). No less than 44 per cent of settlements fall into this category.

The provision of health facilities occurs in different forms, such as mobile clinics, clinics and a hospital. Over 90 per cent of settlements are within 10km of a clinic or clinic stop and within 25km of a hospital. Schools are widely distributed in this area with the result that 83 per cent of settlements are within 5km of a school. The provision of post offices is poor with the only comprehensive facilities being in Ulundi and Babanango. Rural shops provide some postal services and a government programme is being implemented to provide post boxes at central locations such as rural shops and clinics. Similarly, there are few police stations in the area. On the other hand, pension pay-points are widely distributed and these become small markets on pension-pay days. As will be discussed below, pensions (and other government grants) constitute a significant source of income in the area.

Table 15.4 Better-off settlements

Criteria	No. of settlements	%
Relatively better-off (water from a yard connection or borehole less than 1km, pit latrines and electricity)	33	32
Intermediate levels of service (often water 2km or more from settlement)	25	24
Worse-off (water from spring or river, or 2km+, pit latrines, but no electricity)	46	44
Total	104	100

Source: WSDP (2001).

There is no reliable information about the distribution of land line telephones in the area, apart from in the towns of Ulundi and Babanango. On the basis of spot checks during field trips in the area, it appeared that most of it is covered by one of the three mobile phone networks. Circumstantial evidence indicates that there is widespread use of mobile telephones, and it seems likely that at least someone in every settlement would have one. However, sections of some of the main routes leading to the Ulundi-eMakhosini area that are not covered by the national networks, and there are areas within it that have no reception.

15.2.4 The local economy

The main sector contributors to the local Gross Geographic Product (GGP) are government and community services, quarrying and agriculture. The majority of those employed work in the public sector (61 per cent), while 30 per cent work in the private sector, and 9 per cent in the informal sector. Most formal employment is found in government/community services, institutions and private households, as shown in Table 15.5.

The total number of formal jobs (13,000) is, however, far less than the number of people unemployed (25,500). This means that there are almost two people seeking work for every employed person; alternatively, that there is only one local wage paying job for every three households. Herein lies by far the greatest development challenge facing this area. A recent study quoted an employment rate of between 24 per cent and 32 per cent of the economically active population. The remaining 68 per cent to 72 per cent are either unemployed or engaged in informal/rural production activities outside of the formal sector. The main informal sector activities are trade and personal services (Robinson Ellingson Planners, 2000: 7).

Table 15.5 Employment by sector

Sector	%
Farming	5
Mining	4
Manufacture	3
Utilities	<1
Construction	5
Trade	7
Transport	4
Business services	4
Social/community services	38
Private households	9
Institutions	21
Total	100

Source: Demarcation Board (2000).

Table 15.6 Formal sector occupation pattern

Occupation category	%
Management/professional	20
Technical/skilled	11
Clerical/service	29
Craft/machine	17
Elementary/unskilled	23
Total	100

Source: Demarcation Board (2000).

The broad household income distribution in the area reveals two interesting patterns. At the bottom end of the scale are severely impoverished households, who live in the rural settlements. No less than 25 per cent of households reported no income and a further 11 per cent had incomes below R 2,400/month. At the top end of the scale, 11 per cent have incomes in excess of R 30,000/month and 14 per cent between R 12,000 and R 30,000/month (Demarcation Board, 2000, based on 1996 Census). The latter well off groups are mainly urban, but include some people with formal jobs in rural areas, such as teachers, nurses and government officials.

In a recent analysis of rural poverty and livelihoods in South Africa, May (2000) identified six sources from which rural households derive income. The findings which follow are applicable in the area around Ulundi and eMakhosini-Ophathe:

- agriculture for own consumption or sale;
- small and micro-enterprise activities based on the extension of distribution networks, such as hawking, making clothes and handicrafts, child minding, money lending, or contract agricultural services;
- wage labour including migrant labourers, farm workers and commuter labourers. These fall into two sectors – those with secure, well paid jobs with prospects of career advancement and those which are low paid and offer little security or opportunity for upward mobility;
- claiming against the state by way of social pensions and disability grants;
- claiming against household and community members who are working as migrants, in the form of remittances (May, 2000: 25).

The relative importance of these categories is shown in Table 15.7, which also draws attention to the fact that rural households depend on multiple sources of income.

Table 15.7 Income generation and claiming systems

Activity	% households engaging in activity
Agricultural production	36.4
Small and micro-enterprises	10.4
Wage labour in well-paid secure (primary) market	22.1
Wage labour in low-paid, insecure (secondary) market	37.4
Claims against the state (pensions, disability grants, etc.)	32.4
Claims against household members	39.0

Source: May, 2000.

15.2.5 Conclusions

To sum up, the area's fragile economic base is at risk, service levels to rural settlements, where 71 per cent of the population live, are basic at best, with 44 per cent of settlements lacking a combination of the most basic services. There is widespread poverty and very high levels of unemployment. Yet the area has considerable inherent tourism potential. Can this be converted into a sustainable contributor to the local economy in such a way as to address some of these development challenges? The following section explores the nature of the heritage and ecotourism potential.

15.3 Heritage and ecotourism potential

15.3.1 Background

The eMakhosini valley is a microcosm of the history of south-eastern Africa. It is in this valley, at KwaGqokili Hill, just outside the present day town of Ulundi, that the Zulus first defeated their archrivals, the Ndwandwe. This epic battle was to have repercussions for the whole of southeast Africa as Shaka's victory drove out men who would leave their mark on Zimbabwe, Malawi, Tanzania and Mozambique. For these reasons, eMakhosini is considered to be sacred by many Zulu royalists and traditionalists.

Since the late 1980s, the KwaZulu Monuments Council and its successor, Amafa, have been engaged in the conservation and development of eMakhosini in a way that would bring the benefits of tourism without endangering the cultural asset base. The objectives were to:

- conserve ancestral burial places, historical battlefields and other sites of archaeological, historical and cultural significance;
- maintain an ecologically sustainable natural environment which will promote the historical integrity of cultural sites;
- link the historical sites and place them in the context of a thematic cultural-historical tourist attraction.

By the beginning of 2002, Amafa had secured some 14,000ha in the core area, in which most of the Royal graves and important historical sites are located. The eMakhosini is not only a place rich in history and of great natural beauty, but it is also an area of great ecological diversity, ranging from highveld grassland to valley bushveld. These habitats support a variety of wildlife and rare birds.

Nearby, Ophathe Game Reserve was proclaimed in 1991. Situated on the southern banks of the White Mfolozi river, this 8,825ha reserve is less than 10km from Ulundi and, more significantly, at the edge of the eMakhosini valley. The reason for proclamation of Ophathe was 'to serve as a sanctuary for the endangered Black Rhino and possibly other endangered species as well'. Fauna and flora were to be managed and conserved so as to allow sustainable utilization of resources and protection for sensitive ecosystems. In 2001, Ezemvelo KZN Wildlife drew up a Concept Plan for Ophathe Game Reserve which identified zones based on the extent to which the landscape had been modified and linked these to potential tourist activities. It also identified two park development nodes at which a higher density of activities would be concentrated.

15.3.2 A joint approach to development

The notion of developing eMakhosini and Ophathe jointly had been under discussion for several years. In 2001, following agreement that Amafa would become responsible for the management of a piece of intervening land, the two areas became contiguous. This gave impetus to the process, leading to the far-sighted decision by the Ezemvelo KZN Wildlife Executive and Amafa Council to plan and develop the area jointly. It was agreed to name the development 'eMakhosini-Ophathe Heritage Park'; to proclaim it under both Conservation and Heritage legislation; and to produce a joint development plan covering both areas.

These decisions opened the way to achieving a rare combination of opportunities:

- linking a rich historical and cultural heritage site with a Game Reserve;
- assembling a significant area of land (approximately 24,000ha) which can be proclaimed as a single protected area, and with potential for further expansion;
- having a Park that displays significant biodiversity, extending from an altitude of 1,200m above sea level in the west, through the moist mist belt grasslands and ngongoni grasslands, into valley bushveld below 300m;
- providing sufficient area to carry the 'big five' species;
- preserving a culture and history that has already left an indelible mark on the world stage;
- easily accessed by main roads and situated astride one of KwaZulu-Natal's main tourism routes;
- in close proximity to a medium-sized town (Ulundi) and an airport capable of handling large aircraft.

The eMakhosini-Ophathe Heritage Park (EOHP) is the only known example of a combined cultural conservancy and game reserve in Africa. Its development is being planned with a view to application for World Heritage status in due course. The vision adopted by both organizations reads: 'The vision for eMakhosini Heritage Park is to re-create the cultural and natural landscape of nineteenth century KwaZulu, as far as possible, so as to become a premier tourist attraction of world class that operates on a sustainable basis and generates a flow of benefits for local communities.'

However, these remain plans, as yet only at an early stage of implementation.

15.3.3 Other opportunities for heritage and ecotourism in the area

The opportunities for economic development in the area relate to tourism associated with Zulu history and culture, as well as to game reserves. Within a radius of about 60km of Ulundi and the EOHP, there are no less than 40 sites of historical and cultural significance (Robinson et al., 2002: Annexure 5), a major game reserve (the Hluhluwe-Umfolozi Park), three potential community conservation reserves, and a number of well-established private game farms. In addition, there are cultural events such as the well-known Zulu 'reed dance' and an annual schools cultural competition, both of which attract participants and spectators from throughout the province. In addition, there are the Royal Palaces at Nongoma, a well-equipped airport at Ulundi and a 3-star hotel.

In a wider regional context, several tourism routes have been established:

- The Battlefields route has been marketed strongly, but with emphasis on its western southwestern side. Important destinations on this route are Isandlwana and Rorke's Drift, which are within an hour's drive of the EOHP and Ulundi.
- The Rainbow tourism route extends from Mpumalanga through Zululand to Mtunzini on the coast. The initiative is driven by the Publicity or Tourist Associations of the 12 towns along the route.
- The Zululand Birding route is a commercial project that currently markets, develops, and supports birding-based tourism products in the greater Zululand region.

15.3.4 Existing tourist attractions

At present there are four groups of operational tourist attractions which offer a reasonably high standard of service:

1. The KwaZulu cultural museum and historic reserve. The museum, which is located close to Ulundi, focuses on the Nguni-speaking peoples of southeastern Africa, and houses one of the most representative collections of Zulu material in the country, including a famous collection of beadwork. The 300ha historic reserve includes the reconstructed royal residence of King Cetshwayo, which was destroyed by the British, and the site of the last battle of the Anglo-Zulu war of 1879. It also offers accommodation at the uMuzi in traditional 'beehive' huts within a stockade. The head offices of Amafa are in the museum complex (Robinson et al., 2002: 28).
2. Attractions within Ulundi include the Nodwengu site of the grave of King Mpande; the Ulundi battlefield site where the last battle of the Anglo-Zulu war took place in 1879; and tapestries depicting Zulu lifestyle in the Legislative Assembly building.
3. The uMgungundlovu interpretative centre and Zulu King Dingaan's residence, where local tour guides are available.
4. The impressive new Spirit of eMakhosini monument, which also has local tour guides.

In addition, several attractions are currently being upgraded for tourist visits. The sites of three Zulu kings which are located close to uMgungundlovu are being developed to accommodate tourist visits and have local tour guides in attendance. The Ophathe game reserve is in the process of upgrading its internal road network and providing facilities for day visitors. The establishment of a tourist lodge in EOHP is under negotiation and a game stocking programme has commenced.

There is also a proposal to develop a tourist and information centre at 'Ulundi 19', the intersection of the two main routes in the area and at an entrance to the EOHP.

15.4 Translating potentials into viable tourism products: Gaps and obstacles

While these assets constitute a valuable comparative advantage, the fact that only a few have been developed into viable tourism products means that the potential remains unrealized. The question is how to build on the few existing high standard facilities and to translate the potential of some of the others into viable tourism attractions. The thrust of several recent plans in the area has been to establish Ulundi as the 'gateway' to the heart of Zululand, a 'hub from which tourists can gain easy access to this combination of cultural and ecotourism attractions' (Robinson Ellingson Planners, 2000: 49), with eMakhosini-Ophathe Heritage Park as the flagship tourism project (Zululand, 2002; Ulundi, 2002; Iyer Rothaug, 2003). This requires a combination of identification of the appropriate market segments, marketing and development of tourist attractions within the area.

Zululand has appeal in several market segments. These are South African, middle-income tourists interested in historical/cultural attractions or game reserves (or both) looking for clean, safe, affordable facilities and accommodation; South African school children seeking clean, safe exposure to historical, cultural and wild life experiences in low budget group accommodation; upper income South African tourists looking for safe, clean facilities, international class accommodation and game viewing; and finally, international visitors, whose needs closely match those of upper-income South Africans.

The marketing strategy, which is currently more of a written statement than an actively pursued programme, aims to build a brand for Zululand centred around Ulundi and the EOHP, based on:

- a cultural experience centred around the Zulu nation, its history, people, customs and traditions;
- an ecotourism experience centred around game reserves, wildlife and nature (Robinson Ellingson Planners, 2000: 97).

Ulundi is the 'cultural core of the district', around which 'all tourism activity should be focussed on the theme of its being the birthplace of the Zulu nation, incorporating some of the richest historical sites in Africa, wildlife and living cultures' (Iyer Rothaug, 2003: 43).

The main obstacles to tourism development are as follows:

- a lack of developed 'must see' tourism attractions;
- lack of funding and human resources to develop potential attractions into useable products, and to train (and retain) suitable tourism information officers;

- relatively weak marketing and insufficient linking of information about the 'heart' of Zululand to other tourism marketing programmes in Southern Africa;
- lack of accommodation in and around Ulundi to meet the needs of a range of market sectors;
- need for upgrading of two important link roads: the P700 from Ulundi to the Cengeni Gate of Hluhluwe-Umfolozi game reserve; and a section of the R66 from Ulundi through Nongoma to Pongola and the N2 (both are in progress);
- a lack of clean facilities along the main roads leading to Ulundi and the EOHP, and in the town itself;
- perceptions that the Zululand area is not safe for tourists;
- a lack of local awareness about the importance of making tourists feel comfortable and ways of doing this (Robinson Ellingson Planners, 2000; Robinson et al., 2002; Iyer Rothaug, 2003; Schroeder, 2003).

To date, little use has been made of ICT in the development of tourism products or marketing in the area. The Zululand district site (http://www.zululand.org.za) and a more general provincial tourism site (http://www.zulu.org.za) provide some information, while the tourist offices in towns in and around Zululand, and Amafa (which operates the museum and interpretative facilities at the Spirit of eMakhosini and uMgungundhlovu), all respond to telephone enquiries during office hours on weekdays and make reservations. Within the area very few individuals and relatively few businesses have computers and even less have access to the Internet. However, most economically active local people own or have access to a mobile phone, as do all tourist operators and all tourists. In the short term, the best opportunities for promoting tourism using ICT will be via cellular technologies.

15.5 Strategic responses and opportunities for ICT

A local economic development strategy, which is being prepared for Zululand, has formulated six strategies and associated projects in an attempt to address these obstacles and place the 'heart of Zululand' firmly on tourist itineraries (Iyer Rothaug, 2003). These are summarized in Table 15.8. It is interesting to note the emergence of an increasing number of ICT-related measures and projects to develop tourist attractions, drawing attention to a hitherto untapped technology, which could play an increasingly important role in the future.

According to tourism strategist, Lesley Schroeder: 'People have to have a reason for visiting an area in the first place' (Schroeder and Wasserfall, 2003). In this case the anchor attractions are the Museum and cultural centre at Ondini, the Spirit of eMakhosini monument, the interpretative centre (museum) at uMgungundhlovu, and, once developed, the Ophathe game reserve. At this stage, there is a need to communicate these products to the outside world. The starting points, according to Schroeder, should be the development of a focussed product brochure; a tourism information manual; training of tourism information officers;

improving the state of public toilets; a media publicity drive; and improvements to the web site. When tourists start coming, associated services will grow and the benefits will spread locally (Schroeder interview, 2003).

Table 15.8 Tourism elements of the LED Strategy for Zululand in the Ulundi area

Strategies and objectives	Projects
Tourist information inside the region To create a professional information network which uses modern technology and is available 24 hours a day	• Tourism database developed in partnership with a national cell phone network, providing tourists with comprehensive up-to-date information on the region • Physical information offices and tourist refreshment centres on main access routes • Signage throughout the region linked to the promotion of a 24-hour information cell phone number • Information centre on the R34 at the entrance to EOHP ('Ulundi 19')
Tourism information outside the region To make it easier for tourists to find information on tourist attractions in Zululand	• Printed tourism marketing products to be distributed at key locations outside the area, including the main gateways for international tourists at Johannesburg and Cape Town airports • Education programmes for tourist information staff • Upgraded internet information and inclusion of search engines • Media publicity campaign • Education programmes for tourism decision makers • Initiatives by Tourism KZN (the provincial agency) to market the area
Making tourists feel welcome To counter the perceptions that Zululand is not safe and make tourists feel comfortable about travelling in the area	• Ensuring safety of tourists • Community tourism education • Tourism information training for petrol filling station attendants • Marketing improved road linkages
Community tourism To promote the establishment of sustainable community tourism ventures in Zululand by establishing a conducive environment, appropriate supporting structures and guidelines	• Community tourism education, e.g. an awareness street theatre programme to promote the benefits of tourism to communities, and the need to make tourists feel welcome • Tourism training for petrol filling station attendants • Marketing improved road linkages • Clean facilities campaign • Local tourism information centres • Specific community reserves
Visible tourism development To develop a series of 'must see' tourism attractions which	• eMakhosini-Ophathe Heritage Park development • Events around the Royal palaces • Projects in other areas of Zululand (Pongola biosphere,

will draw tourists to the area and simultaneously demonstrate to potential investors that the district is taking tourism development seriously	Vryheid cultural village, Thakazulu and Hot spring development)
Municipal tourism To develop tourist attractions in each of the 5 towns in the district, based on local attributes	• In the case of Ulundi, all tourism activity should be focussed on its being the birthplace of the Zulu nation, proximity to eMakhosini and Ophathe, the airport, and the link road (P700) to Hluhluwe-Umfolozi Park
Institutional structuring for tourism To establish tourism management and implementation structures	• Facilitate private sector representation on district tourism portfolio committees • Development of a district tourism information manual • Improved networking among the key stakeholders in tourism development in both public and private sectors • District and local coordination through the district tourism officer

Source: Iyer Rothaug, 2003: 36–44.

15.6 Conclusions

While the success of the above strategies and projects will only be known in the years to come, it is worth reflecting on the ways in which ICT can add value to tourism as a strategy for regional development in Zululand, and on some broader lessons for other rural tourism initiatives in developing countries. The opportunities for ICT may be viewed from three perspectives: tourists; tourism and travel operators; and the host communities.

Tourists require information before arriving in an area, as well as when they are there, yet according to Myles (quoted in Iver Rothaug, 2003), only 5 per cent physically visit a tourist information office. This points to a significant potential for a regional information office, accessible by cellular phones, operating 24 hours a day, 7 days a week. Schroeder argues the case for erecting large billboards in an area like Zululand advertising the tourist response number. The operators would need to be well-trained and supported by a comprehensive manual covering all the questions tourists might ask. This could provide tourists with information about attractions, accommodation and facilities along the routes; be linked to an accommodation reservation service; track tourists who require someone to guide them to a particular destination; and become a database for subsequent follow-up provision of information (Schroeder interview, 2003).

Travel and tourism operators already make considerable use of the Internet to exchange product and destination information. Similarly, tourism web sites are used extensively by both tourists and by travel and tourism operators. These trends are likely to continue and expand. However, cellular technology, linked to '24/7' tourism information centres, offers a new dimension, given the widespread and rising use of cell phones.

The host communities may be classified into four groups: people working at tourist attractions or facilities (such as information centres, hotels, restaurants); those working in places from which tourists tend to seek information (such as petrol filling station attendants, shop attendants, vendors at roadside stalls); people who can or seek to provide tourists with a particular service; and the host community at large.

For those in the first two categories, a '24/7' cellular network tourism information service could become the basis for local education programmes aimed at enhancing local knowledge about tourist information, as well as a source to which to refer queries. For those offering tourism services, or seeking employment in the tourist sector, a similar call centre could be established on which people register their particular services or products, which might be local crafts, or some local produce. In the same way, people with relevant skills or experience could register their availability, thus providing a type of local employment bureau. A prototype of this was developed in the Cato Manor urban renewal project in Durban (Cato Manor Development Association, 2003: 18). This facility could be established so as to register information in either hard copy or electronic form. Education of the host community can be undertaken via a variety of media.

In conclusion, there appear to be a number of distinct ways in which ICT, and particularly cellular technology can be used to overcome the obstacles to tourism development in the Ulundi-eMakhosini area of Zululand, and (with modification) in other rural areas in developing countries. While ICT may be a vital 'missing ingredient' in the delivery chain, it needs to be used in tandem with simpler, older technologies with which members of most host communities will be familiar.

References

Cato Manor Development Association (2003), *Cato Manor Development Project Review 1994–2002*, CMDA, Durban.

Demarcation Board (2000), *Ulundi Local Municipality Profile*, Demarcation Board web site, <http://www.demarcation.org.za>.

Iyer Rothaug (2003), *Siyaphambili: Zululand Local Economic Development Strategy Report – Draft 2*, Unpublished report to Zululand District Municipality, Ulundi.

May, J. (2000), 'The Structure and Composition of Rural Poverty and Livelihoods in South Africa', in B. Cousins (ed.), *At the Crossroads – Land and Agrarian Reform in South Africa into the 21st Century*, Programme for Land and Agrarian Studies, University of the Western Cape, pp. 21–34.

Robinson, P. in association with planning teams from Amafa and Ezemvelo KZN Wildlife (2002), *eMakhosini-Ophathe Heritage Park Strategic Plan*, Unpublished report to Amafa and Ezemvelo KZN Wildlife, Ulundi.

Robinson Ellingson Planners (2000), *Ulundi Local Economic Development Plan*, Unpublished report to Ulundi Local Municipality, Ulundi.

Schroeder, L. (2003), *Personal communication.*

Ulundi Local Municipality (2002), *Ulundi Integrated Development Plan*, Prepared by Vuka for Ulundi Local Municipality, Ulundi.

Zululand District Municipality (2002), *Water Services Development Plan for Ulundi (DC26)*, Unpublished report to Zululand District Municipality, Ulundi.

Zululand District Municipality (2001), *Zululand Integrated Development Plan*, Prepared by the ZDM technical team, Peter Robinson & Associates and Intermap, Unpublished report to Zululand District Municipality, Ulundi.

Chapter 16

Tourism, Technological Change and Regional Development in Islands

Harry Coccossis

16.1 Introduction

Technological innovation in transport, telecommunications and business organization have, to a great extent, affected not only tourism as a sector but also small tourist destinations, such as small islands, offering new prospects for local and regional development.

There is such a wide diversity of characteristics, conditions and problems that it is doubtful whether one can generalize about islands. In spite of their differences, many islands share at least some common characteristics and problems, such as distance from important markets or centres of activity and decision-making, limited regional natural resources and limited domestic production and consumption systems which condition economic structures and prospects for growth, predominant role of transport in economic exchanges, distinct social and cultural characteristics, etc. (Coccossis, 1987). To what extent these characteristics are dominant depends more or less on size and relative geographic location. Yet it is also the combination of these two factors which makes small islands in peripheral areas particularly interesting cases from the point of view of regional policy.

Small island conditions are often associated with problems of remoteness and isolation, small size of markets and small-scale diseconomies, economic monoculture and dependence on transport, lower relative levels of economic activity and development, etc. Local development opportunities are often restricted by a narrow resource base, small size of local markets and higher costs of access to external markets. This renders small islands more vulnerable to external economic competition which increases the effects and risks from disturbances. It is this combination of isolation and small size which increases the islands' vulnerability to risks of external or internal environmental or socio-economic disturbances.

16.2 The system dynamics of small islands

Small islands are highly dynamic and open systems. As a consequence, they are deeply affected by economic, social and technological change. The impacts of change can be observed in terms of population and societal patterns, economic activity structure and use of resources and, as a consequence, environmental conditions.

Island populations demonstrate intensive fluctuations as a response to economic opportunity. Economic decline (i.e. due to depletion of local resources or natural disasters, etc., or to broader technological, economic, social and institutional changes) has often led to strong outmigration. Economic prosperity for similar reasons (i.e. new resource base, technological improvements, etc.) has led to 'symmetrical' responses of in-migration. Population change, whether by contraction or expansion, might have significant positive or negative impacts on local societies. Migratory patterns affect the structure of local society in a differentiated way, as it is mostly the young and dynamic part of the labour force which is sensitive to economic opportunity shifts. In some cases, it could mean that the local society loses its dynamism, in the form of those who are young, better-skilled and entrepreneurial, thus eroding its future prospects. Whether incoming or outgoing flows are the result, the age-sex societal structure might be affected with eventual side effects on social relations and social life. The ups and downs of population might have an impact on the use of resources as well. Many small islands which are experiencing in-migration present evidence of overutilization of limited-resources and eventual degradation. Similar environmental impacts can be seen from out-migration which could alleviate pressures on limited-resources, but could also be associated with abandonment, ultimately leading to resource degradation. Resource degradation directly or indirectly affects economic (and social) development opportunities, thus triggering further changes in the island system.

Technological change may have profound effects on small islands as it affects accessibility and the costs of production and consumption, and therefore the local economic base and competitiveness. For example, in the past the shift in technology of sea transport from sail to steam has led to the decline of small island economies. In a similar manner, modern transport technology has dramatically improved accessibility and economic and social conditions in many small islands enabling their integration into broader socio-economic systems. Change in telecommunications has also accounted for much change in small islands, influencing production and consumption patterns but mostly services and quality of life. Technological innovation has been beneficial to rural areas allowing access to higher order services obviating to a certain extent the need for personal transfer or reducing the overall costs.

Institutional changes (such as European space integration, sea and air transport deregulation, etc.) are equally important in eventually inducing changes on small islands. Broader institutional changes like the opening up of global markets can widen opportunities but can also marginalize local production systems. Such changes can provide opportunities for the better integration of small island economies in wider economic systems, but can also imply dangers for remote areas in that they become more peripheral and lag behind in development.

The prospects of development for small islands are related to the extent to which they participate in broader networks of interaction and exchange and their capacity to adapt to change. Isolation and remoteness may render them 'immune' to social and economic disturbances in theory, but in an increasingly globalizing world the evidence is the opposite in practice.

16.3 Tourism challenges

Tourism provides a good example of how broader socio-economic, technological and institutional-organizational changes affect many small islands, which often depend on tourism. The main elements that attract tourists in small islands are the relatively unspoilt natural environment, such as the beaches and the sea, the rich and unique landscape, the picturesque villages and local architecture, the local society and lifestyle. An eventual loss or degradation of these assets may also affect tourism itself and ultimately the future of the islands (Coccossis, 2000).

Rising incomes and standards of living, reductions in working hours per week, more time available for leisure, technological changes in air transport, lower costs of travel, and the organization of a travel industry, all account for the substantial increase in tourism after the Second World War. Tourism has become a major source of employment and income for many small islands providing distant places with often much needed opportunities for economic development.

Tourism has many positive impacts on small islands by providing, through cross-linkages, opportunities for other sectors of the economy. Tourism also brings other benefits for local societies through improvements in living conditions, primarily in terms of infrastructure and services (i.e. transport and communications, banking, etc.). Because of the significance of natural and cultural assets for attracting tourism, protecting and enhancing the environment might lead to improving the attraction of a place. Good services, small size, and the quality of the environment and lifestyle can be significant competitive advantages for some small islands in attracting people and economic activities.

Tourism may, however, also include threats to many small islands, thereby affecting social structures and relations, culture and the use of resources in a negative way. Tourism competes with other sectors and increases pressures on local resources. It can lead to monoculture displacing other activities, as income opportunities are in general much greater in tourism, and require much less effort. The higher demand for products from tourism cannot be met by local production systems and this often leads to an increase in imports, which in turn erodes local production patterns. Possible linkages are lost to import substitution. This may lead to the abandonment of production, traditional practices and patterns of cultivation and care of local resources (Coccossis, 1990). Such negative effects may ultimately undermine the prospects for sustainable development in an island in the long term.

Tourism also involves externalities which might be equally important for local societies: for example, rising costs of living, increasing competition with local products, raising expectations, and changing consumption patterns (Lockhart and Drakakis-Smith, 1997). In many tourist destinations, rising costs of living reflect increases in real estate prices, in the basic costs of infrastructure and services, and in the costs of basic products. Such increases may affect disposable incomes of local societies in a negative way, but may also induce social marginalization and displacement, particularly for those social groups which do not necessarily benefit directly from tourism (as, for example, rural households, the elderly, etc.). Competition with local production systems is often in the direction of import substitution, local product displacement, monoculture, and further dependence on imports. Such changes aggravate rural abandonment and the exodus from the rural sector. With increasing opportunities in tourism, expectations rise but this reduces the attractiveness of non-tourism-related sectors and affects their dynamism (in terms of sapping the young and entrepreneurial labour force), thus further increasing dependence. In addition, the adoption of new consumption patterns (towards new products) further increases imports and erodes local economic production systems.

Tourism also affects the natural and man-made environment, which is often the basis for tourist development in many small islands. Tourism consumes local resources (mostly land, water and energy) and generates waste. The disposal of waste may create dangers for the quality of the environment, underground water resources, the beaches, and the sea. Tourism creates other pressures on natural resources as well: for example, by increasing the demand for water and energy which are scarce and inadequate even for domestic uses. Local patterns of water and energy consumption change as standards of living rise due to tourism income. Eventual improvements through technological investment may create spiralling effects by attracting more tourism, which then adds to the burden of local resources and society. There is increasing demand for land for tourist development resulting in higher prices but also increasing the need for public investment. New infrastructure and services improve the attraction of the island leading to further growth of tourism. But this growth threatens the quality of the natural and built environment which is a major asset for tourism. Waste has to be collected, treated and disposed of. The strong seasonal character of peak tourist activity further aggravates building and operating costs of infrastructure. Large-scale investments are required from small island communities to cope with these problems. In addition to these costs of tourism, organizational requirements are high and difficult for small island communities to meet.

Tourism is not the only driving force which creates pressures on small islands. Often the impacts associated with tourism are manifestations of a complex process of development and cannot solely be attributed to tourism alone. Globalization and increasing competition, modernization and technological innovation, geopolitical shifts in international and commercial relations are some of the wider transformations affecting local island systems. However, tourism, as a fast growing activity with strong linkages to several other branches and sectors, is perceived as the major force of change having significant direct and indirect impacts on multiple aspects of island systems: the economy, the local society and the environment. So it is isolated as a major force of change.

Future prospects with respect to tourism growth suggest that destinations such as the small islands are likely to face increasing pressures. For example, in the Mediterranean – which is the primary tourist destination among world regions– international tourism is expected to grow, in spite of the fact that it might lose its proportionate share of world tourism. This does not take into consideration the growth of domestic tourism as well. In addition, as new types of 'selective' tourism (ecotourism, adventure, cultural and other) are likely to account for an increasing share of tourism growth, small islands are likely to increase their attraction and appeal. Their natural and cultural heritage and the 'destination image' which they evoke will be their strong comparative advantage.

16.4 Technological change challenges

Technological and institutional changes are also likely to affect island economies and societies, mainly through the effects such changes imply for the reorganization and competitiveness of tourism destinations. Economic growth prospects are likely to increase as a result of improvements in transport systems (liberalization, lift of cabotage rights, new technology ships, etc.) which will affect access and connectivity, thus reducing travel time and the long-term costs of transport. Such changes are evident in the Aegean islands: in the last decade, the frequency of trips has increased and time distances have been drastically reduced in sea transport, while at the same time the quality of services (better ships with modern facilities) has been dramatically improved. This has also affected tourist flows, bringing in new destinations which in the past were less accessible, so providing outlying undiscovered islands with new opportunities. New infrastructure developments (in terminal facilities and environmental management infrastructures have further increased the attractiveness of existing tourist destinations. Improvements in the technological support systems (air conditioning, telephone and computer services, etc.) of tourist accommodation facilities are helping to attract more and more tourists to small islands. Improvements in telecommunications and information technology (high capacity lines, mobile phone networks, the Internet, etc.) have drastically improved the access of local populations (and tourists) and businesses to better and higher order services (banking, hotel reservation systems, etc.), making small islands attractive for secondary residences and new businesses. The growth of tourism and access to mass media – of various types – have changed attitudes, customs, lifestyles, production and consumption systems, but many of these changes cannot always be assessed as positive. However, human activity systems have also changed as tourism-visitor flows are distributed over longer periods of time, which increases the time-use of island space, thus providing opportunities for better services. Seasonality has improved as a result of more frequent travel trips per visitor per year, a worldwide tendency. Furthermore, the use of islands as secondary residence areas – for the most accessible ones – has brought new life over a longer period of the year in local systems. In the most touristic islands there is an inflow of labour force, while in the less developed ones new opportunities are emerging which reduce tendencies to out-migrate. Better access has in some cases increased seasonal mobility, enhancing flows from the islands to the mainland out of season. People can be located and live in nearby larger islands or urban centres on the mainland in the winter and move back to the small islands in the tourist period. This may reduce out-migration patterns and abandonment, strengthening roots in the islands, in some cases reversing long-lasting trends of population decline. Nevertheless, the diffusion of an 'urban lifestyle' has certainly transformed the structure and dynamics of small islands, bringing them closer to distant areas, thereby integrating them in broader systems of exchange.

16.5 Spatial development: Key policy considerations

Tourism, new technologies and institutional change might account for changing prospects for many small islands, but such benefits are not necessarily automatically widespread. Spatial dynamics intervene to create concentration and diffusion processes which favour some islands over others. So, there is increasing concern about widening intra-regional spatial inequalities even though the diffusion of development opportunities due to tourism, telecommunications and institutional changes beneficial effects on inter-regional differences.

In the last 20 years or so, some island regions, in Greece for example, have demonstrated a reversal in past trends of decline and have even improved their relative position in terms of development, basically as a result of tourism and transport (EC, 1999). In both the Northern and Southern Aegean Regions, the population has grown, and some islands in particular, such as Rhodes and Mykonos, both very touristic, have had population changes several times higher than the regional ones. Economic performance shows similar patterns. So, overall, there is improvement in relative conditions particularly in key tourist destinations. The growth of tourism and transport has also benefited other islands in the regions, as tourism has spread to include them either as new relatively 'virgin and unspoiled' destinations or as overflow from the most frequented ones. Symi, Karpathos, Patmos, Amorgos, Folegandros and several other islands have become increasingly popular with tourists. Tourism growth has spurred the improvement of transport linkages, which benefits other islands as well, bringing them within the tourism circuit, either through the extension of transport services to nearby islands or through the development of local connections as part of extending the tourist services (daily tours, etc.). However, this process is highly selective and has left several other small islands behind in terms of development, which has exacerbated intra-regional inequalities. To some extent, such divergence can be attributed to a lack of strong attractive assets for contemporary tourism (beaches, cultural heritage, etc.), but in many cases it can be attributed to the polarizing effects of the tourist and transport industries. Tourism, particularly the dominant type of mass tourism, is characterized by strong tendencies of vertical integration of services which leads to oligopolistic conditions in the demand side and spatial concentration (Urry, 1990; Shaw and Williams, 1994). Transport systems are by their very nature contributing to this polarization, as they favour certain locations more than others. Improvements in transport facilities and network development can improve an area's attractiveness. In addition, institutional changes in transport (deregulation, etc.) may also contribute to polarization, further widening inter-island disparities even within the same region (Papatheodorou, 2001).

Tourism and technological-organizational change are not the only factors which might influence future development in islands, but as they are fast growing, dynamic and highly interacting, for many islands, they provide a good basis for building a development strategy for their future. Such a strategy should recognize the particularity of islands, focussing on carefully structuring local development options to integrate with regional development strategies. Local development strategies for small islands have to be based on capturing the opportunity to participate in a broad system of spatial interaction on the basis of strengthening their comparative advantages, valorizing endogenous development factors, and respecting the capacity of local systems (including the environment) to sustain growth (Coccossis, 1987). In this context, sustainable development provides a useful general framework of reference, where social equity, economic efficiency and environmental conservation are considered as equally important goals which should be balanced. Spatial integration can be a central axis in a strategy for the sustainable development of islands. Tourism and transport have to be seen as integral parts of such a strategy, as they provide the external stimuli for local growth. Capturing the opportunities offered by tourism and technological-institutional change can be a central objective in the context of a strategy for sustainable development for small islands. However, considering that they are fragile systems, long-term objectives have to be respected in order to reduce the effects of uncertainty and avoid the risks. Size and isolation do not necessarily allow specialization in function, and even if they did, it would not necessarily be in the long-term interest of the islands, because fragility amplifies disturbances to catastrophic proportions (Lanza and Pigliaru, 1995). Tourist development has to be carefully planned, particularly in the case of small islands as the potential negative effects can be disastrous, given the sensitive balance between society, economy and environment (Beller et al., 1990). Protecting and enhancing the natural and cultural heritage can become a strategic element in a sustainable development perspective, where protecting and enhancing the local identity should be the central concern. The strong interdependence of tourism and the quality of the environment (natural and man-made) provides an opportunity to establish such a perspective.

Furthermore, the local development strategies of small islands should be strongly based on local conditions, particularities and constraints in terms of key social, economic and environmental factors and resources, taking into account the phase and type of tourism development. Strategies have to reflect the dynamics of tourism development by specifying development objectives for tourism which can be different for each island. Islands already developed as tourist destinations should adopt tourism development strategies with emphasis on encouraging a reduction and control of tourist activity, while upgrading the quality of services and the environment. On the other hand, islands just emerging as tourist destinations may focus on the development of appropriate planning mechanisms to organize and control the type and scale of future tourist development. The potential impacts of tourist development need to be considered against the stability of the islands' human and natural ecosystems, also giving special attention to the capacity of local systems to cope with tourism from an organizational point of view (Coccossis, 2001).

Small islands have always experienced fluctuations: periods of prosperity and growth and periods of decline and abandonment. To the extent that they have managed to absorb socio-economic-political and environmental shocks with all their destabilizing effects on the fragile balance of economy-society and environment, they have managed to survive and adapt to change. In modern times of globalization, increasing competition, institutional and technological change, with their widespread effects in reducing traditional locational obstacles (such as distance and isolation) and providing small distant places with new opportunities, it is necessary to seek the appropriate mechanisms to capture the potential benefits while reducing possible risks. The challenge for small islands is to use tourism in a broader strategy of long-term sustainable development.

References

Beller, W., P. Hein and P.G. Ayala (1990), *Sustainable Development for Small Islands*, Parthenon Press, Paris.

Coccossis, H. (1987), 'Planning for Islands', *Ekistics*, 54(323/324), pp. 84–7.

Coccossis, H. (1990), 'Historical Land-Use Changes in Mediterranean Europe', in F. Brouwer and M. Chadwick (eds), *Land Use Change in Europe: Processes of Change, Environmental Transformations and Future Patterns*, Kluwer Publishing Co., The Netherlands, pp. 441–61.

Coccossis, H. (2000), 'Sustainable Development for Islands', in G. Maciocco and G. Marchi (eds), *Dimensione Ecologica e Sviluppo Locale: Problemi di Valutazzione*, Franco Angeli, Milano, pp. 237–47.

Coccossis, H. (2001), 'Sustainable Development and Tourism in Small Islands: Some Lessons from Greece', *Anatolia – An International Journal of Tourism and Hospitality Research*, 12(1): 53–8.

European Commission (EC) (1999), *6th Periodic Report on the Socio-Economic Conditions of the Regions EC*, Luxembourg

Lanza, A. and F. Pigliaru (1995), 'Specialization in Tourism: The Case of a Small Open Economy', in H. Coccossis and P. Nijkamp (eds), *Sustainable Tourist Development*, Avebury, Aldershot, UK, pp. 91–104.

Lockhart, D.G. and D. Drakakis-Smith (eds) (1997), *Island Tourism: Trends and Prospects*, Pinter, London.

Papatheodorou, A. (2001), 'Tourism, Transport Geography and Industrial Economics: A Synthesis in the Context of Mediterranean Islands', *Anatolia*, 12(1): 23–34.

Shaw, G. and A.M. Williams (1994), *Critical Issues in Tourism: A Geographical Perspective*, Blackwell, Oxford, UK.

Urry, J. (1990), *The Tourist Gaze: Leisure and Travel in Contemporary Societies*, Sage, London.

Chapter 17

Regional Development, Environment and the Tourist Product

Spyros J. Vliamos

17.1 Introduction

Environmental issues play a very important role nowadays in all aspects of economic policy. Although the Treaty of Rome did not provide the then European Community with the possibility of dealing with these issues, this was remedied by the Single European Act in 1986. Since then, action for environmental protection has been taken in the form of a new legal framework and the introduction of new development policies. This emphasis on environmental institutions was supported by the Maastricht Treaty (Art. 130), the Fifth Environmental Action Programme and the 'Earth Summit's Agenda 21'.

More specifically, the 5th Action Programme was adopted in 1992 and lasted till the end of that decade. It introduced a group of actions aiming at the dissemination of information and the improvement of environmental education of EU citizens. The new framework shifted attention from an independent and partial analysis of several environmental elements (air, water, waste, etc.) towards the formation of an integrated and general management system of these elements, at the level of productive sector and spatial unit. It introduced the provision of what has been called *local conversion plans*, which refer to a simultaneous and integrated development of both the economy and the environment. It also established an interaction among local (economic and social) agents, local authorities and central government by adopting a more global view of environmental problems.

In countries like Greece whose regional development is based to a great extent on the tourism industry and vice versa, the development of a conversion plan would give rise to marketing activities not only for the tourist product but also for the region where the firm operates.

The aim of this chapter is therefore to show the relationship between regional development, environmental quality and the tourist product. The existence of this relation leads to the establishment of a system which makes environment an efficient input to the whole production process.

17.2 Sustainable issues

The perception of sustainable development as introduced by the well-known *Brundtland Report* (WCED, 1987), has become the core concept in any research dealing with development issues. The relationship between *environmental quality* and *economic development* is real and acceptable and has created an ever increasing level of interest, as was shown in the *1992 UN Conference on Environment* and *Development* in Rio (The Earth Summit), and presented in the World Bank Report on the same topic.

This relationship is based on the argument that the quality of the environment has a direct impact on the achievements of the economy. Since the mid-1980s, this notion has always been present in any official document of the European Union (such as the *Fifth Environmental Action Programme of the European Community, 1992*), which stated that, while in the past there has been a tendency for environmental issues to inevitably create a conflict with economic interests, today a reciprocal trend has started to emerge showing a mutual dependence between the two.

An integrated environmental policy may, therefore, contribute to the optimization of resource management and the simultaneous development of some social and economic schemes, which help in creating labour and welfare. European legislation has had an important influence in shaping national environmental policies, as most of the initiatives taken at national levels were based on measures taken by the EU. It can be seen in official documents of the European Community (Single European Act, 1987) that environmental protection measures should be a necessary ingredient of any economic policy and must be based on accurate data for their real impact on the environment to be assessed. Therefore, a *concise, integrated and mutually dependent approach* for the achievement of continuous development should be followed. Something like a 'virtuous cycle' of environment and development has emerged, relating regional development to the environment and local production.

17.3 Sustainable tourism

Tourism has always been regarded as one of the most important factors contributing to the regional, and hence economic, development of a country. Further, it is widely accepted that the tourist industry is influenced by, among other things, the level as well as the trend of the economic activity both worldwide and nationally. This bilateral relationship was observed mainly during the 1990s when the tourist industry faced a major economic crisis, associated with the transition from mass tourism, that is, from a Fordist regime of accumulation, to more customized and individual forms of production, that is, to a Post-Fordist era. This in turn caused a 'territorial reorganization', reflected in the creation of new tourist products and the development of 'new' destinations. This economic restructuring was the reason for the economic decline of the traditional tourist destinations and the tourist industry in general.

This transition from mass tourism has been simultaneously the cause and the result of the dramatic increase in people's interest in the profile of issues relating to the environment, such as *quality* and *sustainability*.

A combination of both the realization of the detrimental impacts of tourism at the destination and the rise of environmentalism and 'green' consciousness in the mid- to late-1980s resulted in a reassessment of the role and value placed on tourism for the development (of some kind) of the tourism destinations. Tourism growth could no longer continue at the present rate without examining the major tourism impacts. In the 1992 *UN Conference on Environment and Development*, travel and tourism was identified as one of the industries which have the potential to make a positive contribution to a healthier planet. Consequently, sustainability issues were established (Agenda 21) and the question became the application of these issues on tourism (sustainable tourism) at national, regional and local levels. Relatively recently, authorities, as a direct result of this new consensus, have started taking an initiating role in the implementation of sustainable tourism practices. However, this implementation process becomes very difficult and hence challenging due to the predominance in the tourism industry of small and heterogeneous businesses.

The basic strategy should aim at the achievement of a convergence of environmental, tourist, and other forms of economic policies, through the participation of the total number of the social and non-social agencies involved (e.g. administration, firms, and citizens). In this way, both goals will be promoted: the coordination of these agencies, on the one hand, and the change of attitudes and the mentalities of the people, on the other.

The unification and the interaction of economic policy and the quality of the environment has a special significance for designing certain forms of policies, since nowadays the quality of life has become an important factor determining the potential of spatial units to attract investment funds. Therefore, the existence of equilibrium between the protection of the natural environment and the level of economic activity is of vital importance for the promotion of a local development policy. It follows that any strategy for local development should combine local society's interests and those of the economic agents (consumers, firms, organizations, etc.). This kind of policy has a severe impact in the shape of a new policy for the development and exploitation of the 'Land of Europe'.

17.4 Environmental improvement at a local level

The *methodology*, which has been proposed to achieve a *better environment and high level of economic activity*, could be the integration factor, which will bring closer heterogeneous (and/or remote) regions. The general approach of this policy lies in the creation of a *strategic local plan*, which will help in developing a comparative advantage for the regions in question, and will be based on a *unified multidimensional environmental management procedure*. The multidimensional character of the plan refers not only to the locality but also to the agents of the economic activity. This strategic plan is known as the *conversion plan*, whose basic element is a *Regional Environmental Management System – REMS*.

The notion of REMS is based on the concept of entrepreneurial viability which, since the 1990s, is the equivalent to quality in every form of economic activity, including environmental quality. Therefore, instead of using a new environmental control procedure accompanied by strict legislation as a counter motive and an additional cost element to the polluters, the authorities may consider applying a REMS in order to convert the *environmental quality level* to an advantage, and, through the introduction of the *regional branding* of the spatial unit in question, gain the economic benefits accrued. The necessary condition for the creation, development and application of REMS is the close cooperation of citizens, firms, public organizations and other economic policy agents. This cooperation is expressed through the development and use of local and regional resources tied up with the development of initiatives taken by those who constitute the indigenous potential of the region. These initiatives will 'stimulate' and cause 'social mobilization' and improve those aspects of development which make use of the existing market forces. In that way, the participation of local firms will become a necessity for them as they will see the benefits arising from this participation. This procedure must begin through information dissemination via vocational training with parallel support from the administration through the provision of consulting services and funds. In fact, there should be a concrete and integrated policy for the region, combining the achievements of environmental goals with the fulfilment of the aspirations of the economic agents.

The application of such a REMS will help the region to acquire 'a competitive edge' leading to a self-reinforced economic development process. The result from the application of such a system to a region is the conversion of the region to a *green region* and some firms in the region can benefit from the adoption of this *green label*. The message to the customers stemming from the firms operating in the region is that the product is produced under strict environmental criteria applied through integrated legal and social procedures.

For this *green label* to be used and introduced to the market, there should be suitable institutions and regulations which will improve and foster the environmental liability of the firms. Additionally, it is a fact that there has been an ever-increasing recognition of *green marketing*, which shifts customers' attention from the product level to the company level and lies behind the whole existence of the latter.

Therefore, REMS is based on three main pillars:

1. initiatives for social mobilization;
2. the dissemination of information;

3. the supply of funds at the local level for the development of new ideas and technologies.

It has been argued (in particular, in the *Fifth Environmental Action Programme of the European Community, 1992*) that local agents who manage the environment can have an important role to play in the creation of those conditions. In that way, the private sector can look at the whole process as an input to its production function and participate enthusiastically, gaining therefore a remarkable share of the benefits from the development. Finally, the goal will be realized through the mechanisms of a social *self-development* process.

It is clear, therefore, that *local development* should be based on the *quality of the environment* as well as the *quality of the product* at every stage of the production process, which, of course, takes place locally. At the product level, emphasis must be given to the distribution network (in the case of the tourism industry by contacts with tour operators, or by the attraction of the individual customer, or in any other form), in relation to the whole product cycle. That is the whole process, which begins with the basic production of the product, proceeds with direct marketing, and ends with the sale of the product.

17.5 An integrated development plan and application strategies

Economic development can be achieved through the development of a local comparative advantage, based on institutions and practices which regulate an integrated environmental management and dictate the upgrading of environmental quality. Therefore, the link between the *company environmental management system* and the relative *regional/local one* becomes a necessity. The two 'parts' can no longer be estimated separately but in combination.

A potential application of this *conversion plan* is the development of a database for the creation of an *Environmental Quality Management Model (EQM)* for the spatial unit in question. The model can be based on a *Geographical Information System (GIS) Model* and can also be used as a database for environmental, social and economic data, using standard criteria with respect to their structure and application.

Finally, REMS must be used as an administrative instrument aiming at the application of a regional integrating plan, similar (or relevant) to the functioning of a corporate environmental administrative system.

As far as the tourism industry is concerned, the planning and management strategies referred to above can be initiated through a *Tourism Development Action Programme (TDAP)*, bearing similar principles and properties to REMS. TDAP aims at the same goals, meant to improve the tourist product and the local business performance and to increase the employment opportunities. However, a number of factors (the tourist industry is strongly influenced by wider macroeconomic forces over which it has little control, lack of available data, or existence of data of limited value, and the potential occurrence of a number of intangible outcomes extremely difficult to quantify) pose several methodological problems and limitations on the application procedure of a TDAP, thus making the assessment of these Programmes a very difficult task.

Therefore for a TDAP to have an effective impact on local and regional development programmes, two potentially important factors must be included:

1. *Cross-organizational cooperation*, within and between the public and private sectors. A successful TDAP forces participants to look beyond their immediate areas, widen their perspectives and access expertise that would not otherwise have been readily available.
2. *Unlimited availability of resources*. In the opposite case, only the pressing issues could be addressed. However, due to the public sector financial restraints, funding has become a serious problem. This can be understood by the fact that there is a constant need for structural reorganization with the application of continuous and innovative regeneration strategies. As pointed out above, tourist behavioural patterns are beginning to display a 'general weariness' with the mass tourism product. The emergence of new destinations and the development of different forms of holidays have served to refocus tourist aspirations and expectations. Although this refocus is due to the fact that any tourism development programme will provide tourism destinations with an initial advantage, which can be copied by competitors and/or become out-dated, it demands resources, which in most of the cases are unavailable.

The application of a TDAP will be supervized by an expert committee as required by the integrated development plan. The committee's duty is to achieve the following strategic goals:

1. The creation and development of an environmental information system, including an environmental monitoring system.
2. The undertaking of an extended socioeconomic and environmental study examining the number of vacancies, any immigration trends and impacts from the development, and infrastructure existing in other regions in proximity.
3. The investigation of the existing communication possibilities among local administrative agents and the corresponding economic ones (industry, commerce, etc.).
4. The creation of internal strategies for the promotion and dissemination of information and vocational training of all the inhabitants with respect to the environmental protection and the benefits accrued from the application of a REMS and/or TDAP.

17.6 Partnerships a way to tourism regional development

Governments in many countries endorse the use of partnership arrangements in planning for tourism development. By encouraging regular, face-to-face meetings among various participants, partnerships have the potential to promote discussion, negotiation, and the building of mutually acceptable proposals about how tourism should develop (Hall, 2000; Healey, 1997). Among the reasons for the growing interest in inter-organizational collaboration is the belief that it may lead to the pooling of knowledge, expertise, capital and other resources, greater coordination of relevant policies, increased acceptance of the resulting policies, and more effective implementation (Pretty, 1995). In such ways, destinations and groups of organizations might gain competitive advantage (Huxham, 1996; Kotler et al., 1993). Further, some commentators contend that there is a moral obligation to involve the range of affected parties in discussions and decisions about potential developments (Innes, 1995; Tacconi and Tisdell, 1992).

Despite their potential advantages, there are often significant difficulties with partnership approaches to planning (Bramwell and Lane, 2000). One potential difficulty is that involving diverse actors in regular meetings and decision-making is usually complex and time-consuming.

Such collaboration can face difficulties because groups refuse to work with others as this may reduce their own influence or power, or because they distrust other parties (Hall and Jenkins, 1995). When stakeholders are involved in joint working, they may not be disposed to listen respectfully to the views of others or to take them into account.

In some places, there may be no tradition of several organizations participating in decision-making. In particular, a participation approach developed in and for developed countries may fail in the socio-economic, cultural, administrative or political circumstances of a less developed country (Roberts and Simpson, 1999; Timothy, 1998, 1999a; Tosun and Jenkins, 1996).

Published work on this theme has grown considerably in recent years, notably research drawing on the collaboration theory (Fyall et al., 2000; Parker, 2000). From this theoretical perspective, joint working can occur when several parties want to respond to a common 'problem domain', but do not individually control enough relevant resources to respond as effectively as they want. Given such 'resource interdependency', several groups may work together if they consider that this would bring them more benefits than acting alone, including through exploitation to obtain resources at the expense of others (Gamm, 1981; Gray, 1989; Selin and Beason, 1991). In this context, Wood and Gray define collaboration as a process where 'a group of autonomous stakeholders of a problem domain engage in an interactive process, using shared rules, norms and structures, to act or decide on issues related to that domain' (1991: 146). The actors are autonomous as they retain independent decision-making powers despite working with each other within a framework of rules or other expectations. Normally, collaborative interaction is considered a relatively formal process involving regular, face-to-face dialogue, these being features that distinguish it from other forms of participation (Carr et al., 1998). Wood and Gray's definition of partnership working also suggests that the people involved must intend to develop a mutual orientation in response to an issue. Gray (1989) argues that collaboration involves joint decision-making among key parties in a problem domain. She identifies five critical features: groups are autonomous but interdependent; solutions emerge by dealing constructively with differences; joint ownership of decisions is involved; groups assume collective responsibility for the future direction of the domain; and collaboration is an emergent process. There may be variations among differing partnerships with these features. For example, they may vary in their duration or time-scale; still, they need to be based on actors interacting on several occasions. Selin (2000) suggests that tourism partnerships intended to promote sustainable development may vary according to such attributes as their geographic scale, legal basis, locus of control, and their organizational diversity and size. Their geographic orientation may be at a community, state, regional, or national scale, while the legal basis for their establishment may come voluntarily from the grassroots or it may be mandated in legislation. Similarly, the locus of control for this shared activity may lie largely with a lead agency or else be dispersed among numerous actors, and their organizational complexity and size may involve just a few groups in only one sector or a large number of parties from multiple sectors. Timothy (1999b) identifies four types of partnerships in the context of tourism planning. As well as the better-known public-private sector form, these partnerships may operate between government agencies, between levels of administration (such as between state and municipality) and between the same administrative level(s) across territorial political boundaries.

There is scope for further research on five aspects of tourism development partnerships related to this study. First, more research could focus on issues involved in joint working at a regional scale, as most previous studies have examined community or local level collaboration (Getz and Jamal, 1994; Jamal and Getz, 1995, 2000; Reed, 2000). Regional tourism planning is taken here to occur at a geographical scale that is subnational but covers a larger spatial area than a local community. Planning for regional tourism is a complex undertaking, representing a formidable challenge for any partnership. A key issue is that it affects multiple groups, such as the government, the private sector, non-governmental organizations, and local communities. These groups will differ according to whether their interests are focussed more at the local, regional, or national scale. For example, the stakeholders with an interest in a region will encompass local municipalities as well as others with interests focussed outside the region, such as the national government and many banks, airlines, hotel chains, and conservation organizations. The complexity of such differences in geographical interest adds to the general difficulty for tourism planning which is the fragmentation of the industry and its interfaces with diverse policy areas, such as transportation, education, and agriculture. But, if regional tourism partnerships are so complex, why might they be useful? In theory at least, they are well-positioned to bring together local, regional, and national interests within a regional development perspective (Inskeep, 1994; Tosun and Jenkins, 1996). They also have the potential to assist national governments to take account of local aspirations and characteristics, and hence reduce tensions among national, regional, and local views. Regional-scale joint activity might also help to secure the goals of redistributing development and related benefits in more equitable ways among more and less developed parts of a country. In addition, such shared arrangements could assist in achieving enhanced coordination among physical, macroeconomic, and social planning, as well as greater cooperation among adjacent local government areas that may have intense political rivalries (Joppe, 1996; Komilis, 1994). Further, the increasing importance of external and often global forces in society may make it logical to plan at a spatial scale larger than local communities, but at a scale that is still 'meaningful' and open to influence by local people (Jenkins, 2000).

A second aspect of tourism partnerships that has received only limited attention concerns the broader contexts within which they emerge, evolve, and eventually are terminated or changed into another form. They are affected by the dynamic interplay between internal and external forces, with the latter including diverse social, cultural, economic, and political influences. Selin and Chavez (1995) tease out some external influences affecting the original setting up of joint working in tourism, describing them as 'antecedents'. These external influences might include a local crisis, the intervention of a convenor, a legal mandate from central government, or prior relations among stakeholders in existing networks. More attention could be paid to how such external influences interact with internal relations in tourism partnerships during their establishment, evolution, and possible closure. Collaborative working in less developed countries may face enormous difficulties from external and internal factors, and these difficulties may be impossible to overcome (Ashley and Roe, 1998; Desai, 1996; Few, 2000). For example, Tosun (2000: 614) identifies a number of 'formidable operational, structural, and cultural limitations' to local communities participating in tourism policy making in developing countries, and these limitations are likely to affect local community participation in partnerships. According to Tosun, one administrative constraint found in many developing countries is that power may be highly centralized in national government, with little remaining with local government. In addition, there are often complex bureaucracies and related jealousies within government in developing countries that fragment the planning process and obstruct coordinated policy making. In some developing countries there is also little experience of democracy, or democracy is largely limited to business, political, or professional elites, or there are clientelistic relations where clients seek favours from powerful patrons. In addition, poor social groups may be uninterested in being involved in planning as they are preoccupied with making ends meet or because of their history of being excluded from decision-making. Further consideration can usefully be paid to how such difficulties affect the scope, if any, for the involvement of local communities and other interest groups in tourism partnerships in less developed countries.

A third aspect of collaborative arrangements ripe for further study concerns the issues they commonly face during their various development phases. Jamal and Getz (1995) advocate the application in tourism research of Gray's (1989) framework of issues that are most common in specific phases of joint working. Table 17.1 summarizes the framework of issues and phases proposed by Gray in 1996. In the first 'problem setting' collaborative phase, the key issues relate to convening relevant stakeholders, securing some common definition of the problem, and getting a commitment to work together. During the 'direction setting' phase, the participants explore the problem in some depth and attempt to reach agreement about a particular direction and related actions. The 'implementation' phase involves ensuring that the agreements reached are followed through into practice, including securing the institutional arrangements for implementation work. While arguing for this general sequence, Gray also emphasizes that 'the phases are not necessarily separate and distinct in practice. Overlapping and recycling back to earlier issues that were not addressed may be necessary' (1996: 62). Gray's framework has been applied in only a few tourism studies. Two applications to this sector are Selin and Chavez's (1995) analysis of three US Forest Service partnerships, and Parker's (2000) examination of planning on the Caribbean island of Bonaire. Both studies provide helpful insights into collaborative working in tourism, indicating the potential value of further applications of Gray's framework.

Concerning the fourth aspect of tourism partnership, further research would be useful on how collaborative work in tourism may incorporate participation by a broader range of interested organizations and individuals. Partnerships usually involve a relatively small number of individuals in regular, face-to-face meetings where they are engaged in shared analysis, and eventually may reach agreements and make decisions together. However, such interorganizational working can be combined with other types of participation involving many more people, with potential to do this through consultation. According to Pretty (1995), consultative participation involves people being asked to express their opinions, and a few stakeholders, usually professionals, listening to the views expressed. While these professionals may modify their decisions in the light of the resulting opinions, they are under no obligation to take these views on board. The professionals retain their power as they define the problems, conduct the information gathering process, control the analysis, and make the decisions about how to respond (Tosun, 1999; Twyman, 2000). The results of such consultation can be reported in partnership meetings, although there is no obligation to heed the views expressed. A variety of techniques can be used to consult with large numbers of people, including drop-in centres, questionnaire surveys, focus group interviews, workshops, and public hearings; further, these techniques can be used in combination (Keogh, 1990; Simmons, 1994; Yuksel et al., 1999). Marien and Pizam argue that 'effective participation programs in tourism ... require a combination of techniques that will work best for its unique set of constituents' (1997: 172).

Table 17.1 Gray's framework of issues and phases in partnership development

Issue	Description
Phase 1: Problem-setting	
Common definition of problem	The problem needs to be important enough to require collaboration and must be common to several stakeholders.
Commitment to collaborate	Stakeholders need to feel that collaborating will solve their own problem. Shared values are key.
Involvement of stakeholders	An inclusive process that includes multiple stakeholders so the problem can be understood.
Legitimacy of stakeholders	Not only expertise but also power relations are important.
Leader's characteristics	Collaborative leadership is key to success. Stakeholders need to perceive the leader as unbiased.
Identification of resources	Funds from government or foundations may be needed for less well-off organizations.
Phase 2: Direction-setting	
Establishing ground rules	Gives stakeholders a sense of fair process and equity of power.
Agenda-setting	Stakeholders' different motivations for joining mean that establishing a common agenda may be difficult.
Organizing subgroups	Large committees may need smaller working groups.
Joint information search	The joint search for information can help to understand other sides of the negotiation and to find a common basis for agreement.
Exploring options	Multiple interests mean that multiple options need to be considered.
Reaching agreement	A commitment is needed to go ahead on a particular course of action.
Phase 3: Implementation	
Dealing with constituencies	Stakeholders need to ensure their constituents understand the trade-offs and support the agreement.
Building external support	Ensuring other organizations that implement are on-side.
Structuring	Voluntary efforts can work, but a formal organization may be needed to co-ordinate long-term collaboration.
Monitoring the agreement and ensuring compliance	This may involve more financial negotiations.

Source: Adapted from Gray, 1996: 61–64.

Finally, with regard to the fifth aspect of tourism partnerships, there will be real value from further evaluations of the extent to which specific partnerships promote participation by local communities in tourism planning. This topic demands further consideration as it is often local people who are left out of the planning and operation of development projects. For example, assessments of tourism partnerships have rarely examined them in relation to the several typologies of local participation devised in other fields of study. While there are many such typologies, 'a basic distinction is that between a system-maintaining and system-transforming process', that is, between local involvement where power and control is held externally as distinct from where it is initiated by local people and they retain control (Shepherd, 1998: 180). Pretty's typology describes seven levels of local participation, ranging from manipulative involvement, where virtually all power and control rest externally with other groups, to self-mobilization, where residents act to change systems by taking initiatives independently of external institutions. The latter does not rule out the involvement of external institutions or advisers, but they are present only as enablers and the local community retains control. The range of forms of involvement in Pretty's typology includes varying degrees of external engagement and local control, and reflects the differing power relations among them. Collaborative arrangements could usefully be examined in relation to this typology.

17.7 The issue of sustainable development

In the general academic literature on sustainable development, much attention has recently been given to the description of different perceptions of sustainable development (Mitlin, 1992; Murdoch, 1993), while attempts to advance the understanding of the concept generally involve discussion of alternative views (Wilbanks, 1994). In exploring the details of the concept of sustainable development, many issues have emerged as points of controversy and departure for adherents to different visions of environmentalism (Table 17.2). These issues have become interwoven in a complex debate on how best to achieve equity of access to natural resources which create human well-being, and on the distribution of the costs and benefits (social, economic, and environmental) which ensue from the utilization of resources (Fox, 1994). Equity implies attempting to meet all basic human needs and, perhaps, the satisfaction of human wants, both now (intra-generational equity) and in the future (inter-generational equity). This means the avoidance of development which maintains, creates, or widens spatial or temporal differences in human well-being.

Under some interpretations of sustainable development, equity is also applied across species barriers, in particular the inherent right of non-humans to exist above and beyond any utilitarian value imposed by humans (Williams, 1994). Far from providing set guidance on the most desirable relationship between the actions of human societies and the status of the natural world, the concept of sustainable development is malleable and can be shaped to fit a spectrum of world-views. These world-views encompass different ethical stances and management strategies and, consequently, range from the extreme resource preservationist stance through to the extreme resource exploitative stance (Turner, 1991).

Table 17.2 Major issues in interpreting sustainable development

1. The role of economic growth in promoting human well-being.
2. The impact and importance of human population growth.
3. The effective existence of environmental limits to growth.
4. The substitutability of natural resources (capital) with human-made capital created through economic growth and technical innovation.
5. The differential interpretation of the criticality of various components of the natural resource base and, therefore, the potential for substitution.
6. The ability of technologies (including management methods such as environmental impact assessment and environmental auditing) to decouple economic growth and unwanted environmental side-effects.
7. The meaning of the value attributed to the natural world and the rights of nonhuman species, sentient or otherwise.
8. The degree to which a systems (ecosystems) perspective should be adopted and the importance of maintaining the functional integrity of ecosystems.

Interpretations of sustainable development can be correspondingly classified as ranging from very strong to very weak (Turner et al., 1994). Rather than detail here all the characteristics which accompany different visions of sustainable development, Table 17.3 summarizes four major sustainable development positions. Some key issues of debate are selected for elaboration.

If there is a growing consensus in defining sustainable development, then it may lie in the frequent rejection of 'extreme' paradigms. Such rejection may result from two factors. The first is an ill-defined perception of a need to become more environmentally-conscious than the traditional resource-exploitative (very weak sustainability) position allows. The second is, in the case of the extreme preservationist (very strong sustainability) position, a feeling that reduced economic activity, population levels, and rejection of much recent technological innovation are so far-reaching as to defy concerted action and, perhaps, some inborn drive in the human psyche. However, alongside these vague notions run compelling arguments for a more central vision of the meaning and implications of sustainable development. Criticism can be directed at both the extreme resource-exploitative and extreme resource-preservationist world-views for effectively ignoring the intra-generational equity principle. With the former, the distribution of socio-economic and environmental development costs and benefits is largely immaterial (Turner, 1991), following a path determined and dedicated, through traditional free market principles, to fulfilling individual (perhaps largely Western) consumer choice. With the latter, the anti-economic growth position appears to deny the world's poor the opportunity of meeting basic needs through economic growth; this position might be described as a new form of (eco)facism (Pepper, 1984). However, a counter argument to this charge can be made on the grounds that economic growth has not helped the great majority of the poor living in less developed countries (Trainer, 1990).

Debate in the 'centre ground' of the sustainability spectrum largely revolves around the compatibility of sustainable development with continued economic growth. Those who advocate a strong interpretation of sustainable development reject the possibility of limitless economic growth (Daly and Cobb, 1989), arguing for a steady-state global economy on the grounds of a perceived need to preserve natural resources and the contribution of ecosystems to maintaining the functional integrity of natural processes (Table 17.3). Here, most, or at least many, natural resources are regarded as critical natural capital, and sustainable development is regarded as requiring adherence to the constant natural assets rule (Pearce et al., 1989), such that the total stock of natural capital assets should remain constant, or rise, through time in terms of quantity and quality.

Table 17.3 A simplified description of the sustainable development spectrum

Sustainability position	Defining characteristics
Very weak	Anthropocentric and utilitarian; growth-oriented and resource exploitative; natural resources utilized at economically optimal rates through unfettered free markets operating to satisfy individual consumer choice; infinite substitution possible between natural and human-made capital; continued well-being assured through economic growth and technical innovation.
Weak	Anthropocentric and utilitarian; resource conservationist; growth is managed and modified; concern for distribution of development costs and benefits through intra- and inter-generational equity; rejection of infinite substitution between natural and human-made capital with recognition of some aspects of natural world as critical capital (e.g. ozone layer, some natural ecosystems); human-made plus natural capital constant or rising through time; decoupling of negative environmental impacts from economic growth.
Strong	(Eco)systems perspective, resource preservationist; recognizes primary value of maintaining the functional integrity of ecosystems over and above secondary value through human resource utilization; interests of the collective given more weight than those of the individual consumer; adherence to intra- and inter-generational equity; decoupling important but alongside a belief in a steady-state economy as a consequence of following the constant natural assets rule; zero economic and human population growth.
Very strong	Bioethical and ecocentric; resource preservationist to the point where utilization of natural resources is minimized; nature's rights or intrinsic value in nature encompassing non-human living organisms and even abiotic elements under a literal interpretation of Gaianism; anti-economic growth and reduced human population.

Source: Adapted from Turner et al., 1994.

For non-renewable resources this implies minimizing loss for future generations through greater efficiency of use, reuse and recycling, where possible. Likewise, the utilization of renewable resources (such as fresh water, soils, natural ecosystems, etc.) should be restricted to operate within the limits imposed by sustainable yield and/or carrying capacity.

Alternatively, for advocates of weak sustainable development (Pezzey, 1989), a greater degree of substitution is possible between natural capital and human-made capital (encompassing economic wealth, built goods, technologies, and the human knowledge base). However, some aspects of the natural environment, such as the stratospheric ozone layer, may be regarded as critical natural capital and, therefore, worthy of absolute preservation. Thus, in general terms, it is enough to maintain, or increase, the total (natural plus human-made) capital stock through time. Essentially, this constitutes a managed and modified global economic growth paradigm, so that economic growth can continue if decoupled from major unwanted environmental side effects via a range of regulatory and market intervention management tools (Table 17.3; Turner et al., 1994).

Something of the contradictory nature of this ongoing debate in the centre ground of the sustainable development spectrum can be appreciated by reference to the perceived role of natural resources in fulfilling the intra-generational equity aim of sustainable development. For example, Pearce et al., (1989) argue that this aim is most likely to be achieved by adherence to the constant natural assets rule for the poor of the developing countries, because in such areas ecosystem productivity is essential to human livelihood, and environmental degradation has a more direct effect on well-being than in developed countries.

Alternatively, Karshenas (1994) uses the link between poverty and environmental degradation to justify a more growth-oriented (weaker) vision of sustainable development, arguing that, below certain levels of economic growth, and in the absence of some required substitution between natural and human-made capital, environmental degradation becomes forced. Certainly, the World Commission (WCED, 1987) has recognized the need for economic growth in poverty-stricken areas of less developed countries in order to meet basic needs. The question of whether or not continued economic growth can be justified in areas of developed countries where such needs are already met and greater well-being largely equates to the satisfaction of wants (for example, more vacation opportunities) has become contentious, at least among academics. The absence of any constraints on continued economic growth in developed countries, under the vision of sustainable development described in the World Commission report, may explain the ease with which sustainable development has become adopted by Western politicians and institutions of economic investment and resource allocation. Thus, the interpretation of sustainable development most frequently presented in governmental policy documents in the West can be described as a weak sustainability position entailing a resource-conservationist, managed growth world-view (Turner, 1991; Wilbanks, 1994). Brundtland (1994), who chaired the 1987 World Commission, has recently endorsed a growth-oriented, weak interpretation of sustainable development by arguing against any suggestion that sustainable development requires a decline in the standard of living for industrialized countries as a means of bringing total global consumption to a sustainable level. However, while flexibility of interpretation may be a source of concern for those who seek or propose a definitive, universally applicable vision of sustainable development for political purposes, others recognize the inevitability of diversity. Indeed, some recent work has highlighted the value which flexibility can bring in terms of understanding human/environment interactions (Healey and Shaw, 1994) with reference to land-use planning and in finding the most appropriate development pathway to follow in often widely different social, economic, and ecological settings (Rees and Williams, 1993). In short, different interpretations of sustainable development will have applicability according to circumstance, involving a different set of trade-off decisions between the various components of sustainability. Turner et al., for example, acknowledge that 'several levels of greenness can coexist within one individual depending on the situation and context under study' (1994: 30).

17.8 The role of tourism in sustainable development

Perhaps the most appropriate way to perceive sustainable tourism is not as a narrowly-defined concept reliant on a search for balance, but rather as an overarching paradigm within which several different development pathways may be legitimized according to circumstance. In other words, there may always be a need to consider factors such as demand, supply, host community needs and desires, and consideration of impacts on environmental resources; but sustainable tourism need not (indeed should not) imply that these often competing aspects are somehow to be balanced. In reality, trade-off decisions taken on a day-to-day basis will almost certainly produce priorities which emerge to skew the destination area-based tourism/environment system in favour of certain aspects. Even over the long term, across several generations, it may be appropriate to abandon any notion of balance in favour of a skewed distribution of priorities. What is crucial, is that tourism development decision-making should be both informed and transparent. In this vein, four possible sustainable tourism approaches, based loosely on the various interpretations of sustainable development, can be outlined and illustrated in an abstract sense. The remit of sustainable tourism is extended to consider the role of tourism in contributing to sustainable development more generally (Hunter, 1997).

1. Sustainable development through a 'Tourism Imperative'
This approach could be seen as going as far as possible towards a very weak interpretation of sustainable development. It is heavily skewed towards the fostering and development of tourism, and would be primarily concerned with satisfying the needs and desires of tourists and tourism operators. The approach

might be most easily justified under three sets of specific circumstances, although tourism would be a new phenomenon, or at least poorly developed, in each scenario: 1) in areas where a strong and demonstrable link exists between poverty and environmental degradation which is characterized by a self-reinforcing cycle; 2) in areas where tourism activity would represent a real improvement upon more overtly degrading current economic activities (e.g. uncontrolled logging, forest clearance for agriculture, or minerals extraction), especially if these bring little benefit to local communities and tourism would create more well-being for more people; and 3) in areas where tourism development would pre-empt the utilization of an area or its resources for other, potentially more degrading, activities. In all three circumstances, tourism could provide the means for some degree of environmental (including local community) protection and environmental education, with tourism-related side effects resulting in the reduced loss of natural resources in terms of quantity and/or quality. However, this interpretation of sustainable tourism might itself involve the substantial loss of natural resources. But, as long as this loss is less than would otherwise occur and does not affect the ability of the area to attract tourists, then circumstances are such that there is no immediate need to aim for tourism development which is particularly sensitive to the local environment, seeks to minimize the consumption of non-renewable resources, or operates within the carrying capacity of local renewable resources (Hunter, 1997).

2. Sustainable development through 'Product-Led Tourism'

This approach may be equated in many ways with a weak interpretation of sustainable development. The environmental side of the tourism/environment system at destination areas may well receive consideration, but this is secondary to the primary need to develop new, and maintain existing, tourism products, with all that this entails in terms of marketing and the encouragement of tourism operators so that growth in the tourism sector can be achieved as far as is feasible. A wide range of environmental and social concerns may be seen as important within the destination area, but, as a general rule, only in so far as these act directly and in an immediately apparent sense to sustain tourism products. This approach might be most easily justified in relatively old and developed tourism enclaves or areas, especially if tourism has come to dominate the local economy. This is because, without the wealth generated by tourism, the well-being of local communities might be compromised to an unacceptable degree, even beginning a spiral of poverty and further environmental degradation (see, for example, English Tourist Board, 1991, with reference to the future of small English seaside resorts). In such places, this alteration of the natural environment may well already be extensive, and attention might focus on actions to beautify the local environment and maintain or improve the built assets (including indirectly supporting infrastructure such as roads or sewage treatment plants) created by utilizing natural resources. These actions might then enable new, perhaps more 'up-market', tourism products to be developed, given an appropriate development climate created through public/private sector partnerships. Illustrative examples of this perception of sustainable tourism can be found in the case studies of Majorca (Morgan, 1991), Conwy in Wales (Owen et al., 1993), and Languedoc-Roussillon in France (Klemm, 1992). If tourism can be sustained in this manner in specific locations, through a primarily economic approach, this might carry the additional benefit of avoiding tourism-related damage to nearby pristine locations where tourism is either unwanted by locals and/or would occur in environmentally-sensitive locations. Thus, environmental protection within a wider geographical region may actually depend on the continued success of tourism in popular and well-defined tourism centres. However, an enlightened vision of tourism development under this approach might also see national or local government authorities earmarking some portion of tourism revenues to decouple the local economy slowly from reliance on tourism towards greater diversity and, therefore, resilience to external change; a position in keeping with many interpretations of sustainable development more generally (Hunter, 1997).

3. Sustainable development through 'Environment-Led Tourism'

In this approach, decisions are made which skew the tourism/environment system towards a paramount concern for the status of the environment. Perhaps most applicable in areas where tourism is non-existent or relatively new, its aim would be to promote types of tourism (e.g. ecotourism, but as more than a mere label) which specifically and overtly rely on the maintenance of a high quality natural environment and/or cultural experiences. The goal would be to make the link between tourism success and environmental quality so strong that it is transparent to all interested parties what the risks to tourism's continued survival would be if tourism were not strictly controlled and ultimately limited to within the carrying capacity or sustainable yield of the least robust aspect of the environmental resource base. Arguably, there is still a very strong product focus with this approach, but it differs from 'product-led' (Scenario 2) tourism in prioritizing environmental concerns over marketing opportunities. An example which reflects some aspects of the environment-led interpretation of sustainable tourism can be found in the work of Sanson (1994) with respect to ecotourism in sub-Antarctic islands. However, the environment-led approach must entail more than finding the most appropriate form of tourism for a particular area, or even zoning to limit access to some sites. Under this interpretation of sustainable tourism, opportunities to work in harmony with, rather than compete against or exclude, other locally-important economic sectors would be seized, as would opportunities to create touristic experiences which highlight environmentally-conscious living through reducing impacts on the immediate and wider environment. Small tourism centres might, for example, promote themselves on the basis of the efficient use of water and energy resources and through materials recycling. Conceivably, environment-led tourism may also be an appropriate strategy in larger, developed centres or areas seeking a new market niche, where promotion is based on genuine and visible attempts to reorient tourism activities along more ecocentric lines.

Reorientation might be attempted through a variety of regulatory and/or market-based techniques designed as 'sticks and carrots' to change the behaviour of tourism operators and tourists themselves. The importance of regulation and encouraging the greater use of waste-free and low-waste technologies by tourism businesses has recently been highlighted by Lukashina et al. (1996), in the more environmentally-conscious development of the Sochi region of the Black Sea coast in Russia (Hunter, 1997).

4. Sustainable development through 'Neotenous Tourism'
This, very strong, sustainability approach is predicated on the belief that there are circumstances in which tourism should be actively and continuously discouraged on ecological grounds. In some places, including nature reserves of national or international importance, tourism growth should be sacrificed for the greater good. Tourism can never be totally without environmental impacts, but one can take the precautionary approach to environmental protection to a point where the functional integrity of natural ecosystems at the destination area as a whole is protected as far as is feasible. Absolute preservation may also be possible at some exceptionally sensitive sites in the sense of maintaining an ecologically viable range of habitats and species (see, for example, Hall and Wouters, 1994, with respect to nature tourism in the sub-Antarctic).

Clearly, 'neotenous tourism' could only apply in areas largely devoid of tourism activity. The word 'neotenous' implies that tourism activities would be limited to the very early, juvenile, stages of tourism development through, for example, the use of permits for access, or through land-use planning development control to prevent the expansion of tourism-related infrastructure. Borrowing from Butler's (1980) tourist-area cycle of evolution, the aim would be to keep tourism development to exploration or involvement stages, perhaps dominated by a small number of individual adventure travellers, small groups of tourists, or those engaged in legitimate study. Likewise, in keeping with a very strong sustainability position, the aim would be to minimize the utilization of renewable and non-renewable resources within these areas (Hunter, 1997).

References

Ashley, C. and D. Roe (1998), 'Enhancing Community Involvement in Wildlife Tourism: Issues and Challenges', *Wildlife and Development Series 11*, International Institute for Environment and Development, London.
Bramwell, B. and B. Lane (2000), 'Collaboration and Partnerships in Tourism Planning', in B. Bramwell and B. Lane (eds), *Tourism Collaboration and Partnerships: Politics, Practice and Sustainability*, Channel View, Clevedon, pp. 1–19.
Brundtland, G.H. (1994), 'The Challenge of Sustainable Production and Consumption Patterns', *Natural Resources Forum*, 18: 243–6.
Butler, R.W. (1980), 'The Concept of a Tourist-Area Cycle of Evolution and Implications for Management', *The Canadian Geographer*, 24: 5–12.

Carr, D., S. Selin and M. Schuett (1998), 'Managing Public Forests: Understanding the Role of Collaborative Planning', *Environmental Management*, 22: 767–76.

Daly, H. and J.B. Cobb (1989), *For the Common Good: Restructuring the Economy Toward Community, the Environment and a Sustainable Future*, Beacon Press, Boston, MA.

Desai, V. (1996), 'Access to Power and Participation', *Third World Planning Review*, 18: 217–42.

English Tourist Board (1991), *The Future of England's Smaller Seaside Resorts: Summary Report*, ETB, London.

Few, R. (2000), 'Conservation, Participation and Power: Protected-Area Planning in the Coastal Zone of Belize', *Journal of Planning Education and Research*, 19: 401–8.

Fox, W. (1994), 'Ecophilosophy and Science', *The Environmentalist*, 14: 207–13.

Fyall, A., B. Oakley and A. Weiss (2000), 'Theoretical Perspectives Applied to Inter-organizational Collaboration on Britain's Inland Waterways', *International Journal of Hospitality and Tourism Administration*, 1: 89–112.

Gamm, L. (1981), 'An Introduction to Research in Inter-organizational Relations', *Journal of Voluntary Action Research*, 10: 18–52.

Getz, D. and T. Jamal (1994), 'The Environment-Community Symbiosis: A Case for Collaborative Tourism Planning', *Journal of Sustainable Tourism*, 2: 152–73.

Gray, B. (1989), *Collaborating: Finding Common Ground for Multi-Party Problems*, Jossey-Bass, San Francisco.

Gray, B. (1996), 'Cross-Sectoral Partners: Collaborative Alliances Among Business, Government and Communities', in C. Huxham (ed.), *Creating Collaborative Advantage*, Sage, London, pp. 57–79.

Hall, C. (2000), *Tourism Planning: Policies, Processes and Relationships*, Prentice Hall, Harlow.

Hall, C. and J. Jenkins (1995), *Tourism and Public Policy*, Sage, London.

Hall, M. and M. Wouters (1994), 'Managing Nature Tourism in the Sub-Antarctic', *Annals of Tourism Research*, 21: 355–474.

Healey, P. (1997), *Collaborative Planning: Shaping Places in Fragmented Societies*, Macmillan, London.

Healey, P. and T. Shaw (1994), 'Changing Meanings of "Environment" in the British Planning System', *Transaction of the Institute of British Geographers*, 19: 425–38.

Hunter, C. (1997), 'Sustainable Tourism as Adaptive Paradigm', *Annals of Tourism Research*, 24(4): 850–867.

Huxham, C. (1996), 'The Search for Collaborative Advantage', in C. Huxman (ed.), *Creating Collaborative Advantage*, Sage, London, pp. 176–80.

Innes, J. (1995), 'Planning Theory's Emerging Paradigm: Communicative Action and Interpretive Practice', *Journal of Planning Education and Research*, 14: 183–90.

Inskeep, E. (1994), *National and Regional Tourism Planning: Methodologies and Case Studies*, Routledge, London.

Jamal, T. and D. Getz (1995), 'Collaboration Theory and Community Tourism Planning', *Annals of Tourism Research*, 22: 186–204.

Jamal, T. and D. Getz (2000), 'Community Roundtables for Tourism-Related Conflicts: The Dialectics of Consensus and Process Structures', in B. Bramwell and B. Lane (eds), *Tourism Collaboration and Partnerships: Politics, Practice and Sustainability*, Channel View, Clevedon, pp. 159–82.

Jenkins, J. (2000), 'The Dynamics of Regional Tourism Organizations in New South Wales, Australia: History, Structures and Operations', *Current Issues in Tourism*, 3: 175–203.

Joppe, M. (1996), 'Sustainable Community Tourism Development Revisited', *Tourism Management*, 17: 475–9.

Karshensas, M. (1994), 'Environment, Technology and Employment: Towards a New Definition of Sustainable Development', *Development and Change*, 25: 723–56.

Keogh, B. (1990), 'Public Participation in Community Tourism Planning', *Annals of Tourism Research*, 17: 449–65.

Klemm, M. (1992), 'Sustainable Tourism Development: Languedoc-Roussillon Thirty Years On', *Tourism Management*, 13: 169–80.

Komilis, P. (1994), 'Tourism and Sustainable Regional Development', in A. Seaton (ed.), *Tourism: The State of the Art*, Wiley, Chichester, pp. 65–73.

Kotler, P., D. Haider and I. Rein (1993), *Marketing Places: Attracting Investment, Industry and Tourism to Cities, States and Nations*, Free Press, New York.

Lukashina, N., M. Amirkhanov, V. Anisimov and A. Trunev (1996), 'Tourism and Environmental Degradation in Sochi, Russia', *Annals of Tourism Research*, 23: 654–65.

Marien, C. and A. Pizam (1997), 'Implementing Sustainable Tourism Development Through Citizen Participation in the Planning Process', in S. Wahab and J. Pigram (eds), *Tourism, Development and Growth: The Challenge of Sustainability*, Routledge, London, pp. 164–78.

Mitlin, D. (1992), 'Sustainable Development: A Guide to the Literature', *Environment and Urbanization*, 4: 111–24.

Morgan, M. (1991), 'Dressing up to Survive: Marketing Majorca Anew', *Tourism Management*, 12: 15–20.

Murdoch, J. (1993), 'Sustainable Rural Development: Towards a Research Agenda', *Geoforum*, 24: 225–41.

Owen, R.E., S. Witt and S. Gammon (1993), 'Sustainable Tourism Development in Wales: From Theory to Practice', *Tourism Management*, 14: 463–74.

Parker, S. (2000), 'Collaboration on Tourism Policy Making: Environmental and Commercial Sustainability on Bonaire, NA', in B. Bramwell and B. Lane (eds), *Tourism Collaboration and Partnerships: Politics, Practice and Sustainability*, Channel View, Clevedon, pp. 78–97.

Pearce, D., A. Markandva and E.B. Barbier (1989), *Blueprint for a Green Economy*, Earthscan, London.

Pepper, D. (1984), *The Roots of Modern Environmentalism*, Croom Helm, London.

Pezzey, J. (1989), 'Economic Analysis of Sustainable Growth and Sustainable Development', *Environment Department Working Paper 15*, World Bank, Washington DC.

Pretty, J. (1995), 'The Many Interpretations of Participation', *In Focus*, 16: 4–5.

Reed, M. (2000), 'Collaborative Tourism Planning as Adaptive Experiments in Emergent Tourism Settings', in B. Bramwell and B. Lane (eds), *Tourism Collaboration and Partnerships: Politics, Practice and Sustainability*, Channel View, Clevedon, pp. 247–71.

Rees, I. and S. Williams (1993), *Water for Life: Strategies for Sustainable Water Resource Management*, Council for the Protection of Rural England, London.

Roberts, L. and F. Simpson (1999), 'Developing Partnership Approaches to Tourism in Central and Eastern Europe', *Journal of Sustainable Tourism*, 7: 314–30.

Sanson, L. (1994), 'An Ecotourism Case Study in Sub-Antarctic Islands', *Annals of Tourism Research*, 21: 344–54.

Selin, S. (2000), 'Developing a Typology of Sustainable Tourism Partnerships', in B. Bramwell and B. Lane (eds), *Tourism Collaboration and Partnerships: Politics, Practice and Sustainability*, Channel View, Clevedon, pp. 129–42.

Selin, S. and K. Beason (1991), 'Inter-organizational Relations in Tourism', *Annals of Tourism Research*, 18: 639–52.

Selin, S. and D. Chavez (1995), 'Developing an Evolutionary Tourism Partnership Model', *Annals of Tourism Research*, 22: 844–56.

Shepherd, A. (1998), *Sustainable Rural Development*, Macmillan, London.

Simmons, D. (1994), 'Community Participation in Tourism Planning', *Tourism Management*, 15: 98–108.

Tacconi, L. and C. Tisdell (1992), 'Rural Development Projects in LDCs: Appraisal, Participation and Sustainability', *Public Administration and Development*, 12: 267–78.

Timothy, D. (1998), 'Cooperative Tourism Planning in a Developing Destination', *Journal of Sustainable Tourism*, 6: 52–68.

Timothy, D. (1999a), 'Participatory Planning: A View of Tourism in Indonesia', *Annals of Tourism Research*, 26: 371–91.

Timothy, D. (1999b), 'Cross-Border Partnership in Tourism Resource Management: International Parks Along the US-Canada Border', *Journal of Sustainable Tourism*, 7: 182–205.

Tosun, C. (1999), 'Towards a Typology of Community Participation in the Tourism Development Process', *Anatolia: An International Journal of Tourism and Hospitality Research*, 10: 113–34.

Tosun, C. (2000), 'Limits to Community Participation in the Tourism Development Process in Developing Countries', *Tourism Management*, 21: 613–33.

Tosun, C. and C. Jenkins (1998), 'Regional Planning Approaches to Tourism Development: The Case of Turkey', *Tourism Management*, 17: 519–32.

Trainer, T. (1990), 'A Rejection of the Brundtland Report', *Dossier 77*, IFAD, Washington DC.

Turner, R.K. (1991), 'Environment, Economics and Ethics', in D. Pearce (ed.), *Blueprint 2: Greening the World Economy*, Earthscan, London, pp. 209–24.

Turner, R.K., D. Pearce and I. Bateman (1994), *Environmental Economics: An Elementary Introduction*, Harvester Wheatsheaf, Hemel Hempstead.

Twyman, C. (2000), 'Participatory Conservation? Community-Based Natural Resource Management in Botswana', *Geographical Journal*, 166: 323–35.

Wilbanks, T.J. (1994), '"Sustainable Development" in Geographic Perspective', *Annals of the Association of American Geographers*, 84: 541–56.

Williams, C. (1994), 'From Red to Green: Towards a New Antithesis to Canitalism', in G. Haughton and C. Williams (eds), *Perspectives Towards Sustainable Environmental Development*, Avebury, Aldershot, pp. 165–80.

Wood, D. and B. Gray (1991), 'Toward a Comprehensive Theory of Collaboration', *Journal of Applied Behavioural Science*, 27: 139–62.

World Commission on Environment and Development (1987), *Our Common Future*, Oxford University Press, Oxford.

Yuksel, F., B. Bramwell and A. Yuksel (1999), 'Stakeholder Interviews and Tourism Planning at Pamukkale, Turkey', *Tourism Management*, 20: 351–60.

Index